Computational Chemistry

Computational Chemistry

Jeremy Harvey

OXFORD
UNIVERSITY PRESS

Great Clarendon Street, Oxford, OX2 6DP,
United Kingdom

Oxford University Press is a department of the University of Oxford.
It furthers the University's objective of excellence in research, scholarship,
and education by publishing worldwide. Oxford is a registered trade mark of
Oxford University Press in the UK and in certain other countries

Published in the United States of America by Oxford University Press
198 Madison Avenue, New York, NY 10016, United States of America

British Library Cataloguing in Publication Data
Data available

Library of Congress Control Number: 2017954085

ISBN 978-0-19-875550-0

Printed and bound by
CPI Group (UK) Ltd, Croydon, CR0 4YY

Preface

Computational Chemistry has recently become an essential research method in chemistry as well as in related areas of the molecular sciences, and is increasingly present in undergraduate curricula. This increasing importance has been enabled by the huge increase in computer power, but also by the development of new methods and of new efficient computer programs that implement them. There are many excellent textbooks dedicated to the subject—or to its theoretical underpinnings—but these can be daunting by their length or their level. This book is designed to provide a concise and accessible introduction to all the key methods and techniques of the field, including quantum chemical electronic structure methods, molecular mechanics, geometry optimization approaches, molecular simulation techniques, statistical mechanics, and hybrid methods. The emphasis is not on providing a systematic description of theoretical chemistry, but instead on providing a few key explanations concerning how each method works. A deliberate attempt has been made to keep the mathematical content low.

The book is primarily aimed at advanced undergraduate students, though it could also be of value to more advanced students and researchers seeking to carry out research in computational chemistry. Also, almost every research chemist will now encounter computational work in the literature relevant to their topic, and the book could provide assistance to such readers both in identifying the methods being used and in assessing the solidity of the conclusions.

While writing the book, a number of illustrative calculations have been carried out by its author, and described in some detail, to help provide insight into how calculations are performed in practical terms. A very wide range of free and commercial computational software is available. The book is not written with the users of any particular program in mind, but tries to emphasize common aspects to all of them. Computational chemistry is in some ways best learned by practising, and as well as repeating some of the calculations described in the book (sample input files for many of them are provided separately), the reader is encouraged to undertake his or her own mini-research projects, in the form of the exercises suggested at the end of each chapter.

The author has been privileged to learn about computation from many colleagues, students, and mentors, and would like in particular to acknowledge the insights received from V. Aggarwal, M. Ashfold, N. Fey, F. Manby, A. Mulholland, and A. Orr-Ewing (Bristol), R. B. Gerber (Jerusalem), D. Schröder (formerly in Prague), M. Aschi (l'Aquila), and A. Ceulemans (Leuven).

Leuven
July 2017
J.N.H.

Acknowledgements

The author and publisher would like to thank the following individuals for providing valuable feedback on draft chapters during the writing process:

David Benoit, University of Hull
Furio Cora, University College London
Jonathan Hirst, University of Nottingham
Syma Khalid, University of Southampton
Alessandro Troisi, University of Warwick
Herrebout Wouter, University of Antwerp

Contents

1 Computation and Computers in Chemistry

1.1 Introduction

Computation in the sense of calculating something or other is ubiquitous in chemistry. Devising an experiment or analysing its results usually involves some calculation or other—if only a simple one, such as the sum required to work out how many grams of reagent you need to make up 30 mmol. A more fundamental role for computation is to work out the details of the prediction of a particular theory. Very often, this requires so many calculations that they cannot be done by hand or with a calculator—a computer is needed instead.

In modern chemical research, computation is used for such diverse aims as studying the diffusion of molecules through a membrane, for exploring the motions of one part of a protein with respect to the others, for predicting the features in an electronic absorption spectrum, or for trying to understand the mechanism of a chemical reaction. The aim of this book is to explore how the corresponding methods work and to introduce the reader to the techniques used to carry out such calculations.

1.2 Theories and computation in chemistry

Consider a theory that predicts the appearance of the infrared absorption spectrum of a diatomic molecule in the gas phase from knowledge of the masses m_1 and m_2 of the two atoms involved, and the shape of the potential energy curve for interaction between the two atoms. This theory can be written in the form of an equation for the *vibrational frequency* v:

$$v = \frac{1}{2\pi}\sqrt{\frac{k}{\mu}} \tag{1.1}$$

where μ is the *reduced mass* of the two atoms (and is equal to $m_1 m_2 /(m_1 + m_2)$), and k is the *force constant* of the bond, with the potential energy V for the system assumed to be given by equation (1.2), for values of the interatomic distance r close to r_0, the equilibrium bond length:

$$V = \frac{1}{2}k(r - r_0)^2 \tag{1.2}$$

This theory can be tested and confronted with experiment in a variety of ways that will require some calculations to be performed. One possibility is that the frequency v can

Potential energy curve

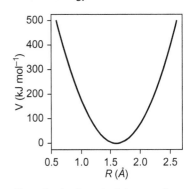

Curve showing the potential energy of a diatomic molecule as a function of the distance between the two atoms.

be measured experimentally for a given molecule, such as the major isotopomer of hydrogen chloride, $^1H^{35}Cl$. Then you can calculate the value of μ for this pair of atoms, and invert the equation above to obtain a value for k. Then, you can use the equation as written, but now using a modified reduced mass, e.g. that for a different isotopomer, such as $^2H^{35}Cl$ or $^1H^{37}Cl$. You can then compare the prediction to experiment. For this simple example, there is no real need of a computer, since the equation that expresses the theory is very simple, and the necessary calculations can be evaluated by hand (or with a calculator).

However, it is not all that difficult to come up with a very closely related situation in which the calculations are much harder to carry out by hand. For example, you might decide to go beyond the approximate second-order Taylor expansion of the potential energy surface of equation (1.2) to a more accurate expression, such as a Morse curve potential or indeed a more complex one. Calculating V for a given distance r then becomes harder, especially if you wish to repeat the calculation for many different values of r. Even more challenging is to carry out a similar analysis for a polyatomic molecule—such as chlorobenzene C_6H_5Cl, with its twelve atoms and 30 internal degrees of freedom. It is possible to write down expressions for the potential energy V in such systems based on the various bond lengths and angles in the molecule (see Chapter 4) or to calculate V for given arrangements of the atoms based on the laws of quantum mechanics (see Chapter 2). It is also possible to work out the form of the theory that relates the values of V to the vibrational frequencies (see Chapter 5). However, going from the experimental data to extract the force constants, or from the force constants to the vibrational frequencies, is now quite a complicated procedure to carry out by hand or even with a calculator, and in most cases will instead need to be done with a computer. And this is still a very simple molecule! If you want to do a similar analysis for a protein surrounded by solvent, with many thousands of atoms, a computer is quite simply mandatory.

Many problems of this type occur throughout chemistry: you have a theory that accounts for the value of some observable in terms of other properties of the system, but this theory is so complicated that its predictions cannot, in practice, be worked out using only paper and pencil. These are the cases where computational approaches are needed. Computers are very good at carrying out calculations that are much too demanding for paper and pencil, with millions of repeated operations.

In this description, the computer does not, however, do all the work for the scientist: he or she still needs to choose or develop a theory that might account for the observable property that is to be studied. This involves choosing the background framework within which to describe the system, and also choosing a set of approximations that lead to a model that is—hopefully—accurate enough, yet is also tractable for a computer. Deriving a theory from some other more fundamental theory or concept may sound like something that a computer would do well, since it will involve mathematics. However, generally speaking, computers are not very good at mathematical derivations. They excel at repeated calculations, but do not straightforwardly carry out derivations or any other operation that involves *symbolic* manipulation of functions or other abstract mathematical objects.

It should also be noted that computers do not completely remove another very important part of the scientist's work: assessing the level of agreement between the outcome of the computations based on the theory, and the experimental observations. In some cases, computed data such as statistical measures of the deviation between experiment and predictions will be used as part of the assessment. But the computer cannot carry out the final judgement.

1.3 How computers work

To understand computational methods in chemistry, it is helpful to have a schematic understanding of how computers work. At the heart of a computer is the central processing unit, or CPU. This carries out operations on numbers: at the most fundamental level, the CPU contains a set of transistors which take in two input signals—a current that is either on ('1') or off ('0') and outputs a current that is itself either on or off. This extremely simple behaviour is equivalent to carrying out logical operations. The CPU is set up to handle many such operations on bits of data (a '0' or '1') at the same time— e.g. 32 or 64 bits can be handled at one time. By combining the corresponding output signals, the CPU is able to carry out a more sophisticated arithmetic operation such as adding, subtracting, multiplying, or dividing two integers or decimal numbers. More sophisticated operations like evaluating a trigonometric function, or a logarithm, may take more than one step. Modern CPUs are able to perform many sets of operations (or *cycles*) every second—this is the frequently quoted clock speed of the CPU, a measure of how many cycles of operations the CPU can perform each second. A 3 GHz processor core, fairly typical nowadays, can carry out roughly *3,000,000,000* multiplications per *second*. Modern computers also typically contain CPUs that are able to carry out multiple operations simultaneously, or in parallel—these CPUs have multiple cores.

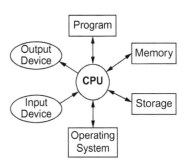

Which operations does the CPU carry out? You might imagine that it performs sums that have been requested by the user, who has entered the request using an appropriate input device—typically a keyboard, but it could also be a mouse, another computer via a network, or some other thing. The result of the operation would then be sent to an appropriate output device—a screen, a printer, another computer, or such like. This is roughly what a calculator does, but computers tend to work in a somewhat different way. The CPU carries out a sequence of operations that are dictated to it by a program. The program tells the CPU to take input for a given operation from either an input device, or an internal source. The output from the operation is sent either to an output device or an internal destination. The instructions that cause these things to happen are themselves operations on bits of data. Broadly speaking, the internal places where data is stored can be split into memory and storage (a hard disk, or solid-state drive), with memory being much faster than storage at communicating with the CPU, but having a lower capacity.

What gives the program the 'authority' over the CPU? This is the role of the operating system, which is a form of over-arching program that runs on the computer from the moment it is switched on. It processes input from some or other input device that requests that a given program should execute, instructs the CPU to read the set of instructions that make up the program from a source, such as the storage device, and then makes the CPU carry out those instructions.

In computational chemistry, most programs have, as their main function, to carry out a large set of calculations, as described previously. The set of calculations to be performed is specified either within the program itself, or in a set of additional instructions that can be input by the user. Most often, the latter instructions are conveyed to the computer either by typing in some commands at the keyboard, through use of some

Algorithm Sequence of mathematical operations that together provide the result to a given problem.

other input device, or through one or more input files. Most often, input files are made up of plain text—letters and numbers, which may specify things like the number and nature of the atoms in the system to be modelled, the initial positions of these atoms to be considered, and the details of the calculation to be performed. The program will then start to run (or 'execute') and then the results are output. Usually this is done through one or more output files, usually also in plain text, that are written to the storage device, and can later be read by the user through printing or display on screen. In many cases, generation of the input and analysis of the output is facilitated by using an auxiliary program that can interconvert between the plain text input and output and a graphical representation. This program is called a graphical user interface (or GUI). In some programs, especially those for casual users, the GUI will be combined with the main computing part to form a single program.

Most of the calculations carried out by computational chemists use pre-written programs written by other scientists, that convert an abstract algorithm into a set of instructions for the CPU. This is almost always done in some form of high-level programming language, such as Fortran, C++, or Python. The instructions, as written in these languages, map fairly naturally onto the set of equations that are included in the theory to be tested. For example, you may write a command that requests that the cosine of a large number of angles be expressed, with the result for each angle being multiplied by a certain number, and the result being assigned to some temporarily stored location, where it can be used thereafter for some other operation. This whole instruction could be written in some form like 'V = c* cos(theta)'. Some programs can be executed directly—with the high-level instructions being converted (by some auxiliary program) into detailed instructions to the CPU to carry out a sequence of elementary operations while the program runs. In other cases, the program expressed in the high-level language will first be converted into a set of detailed instructions for the CPU—a version of the program in a low-level language. This conversion is carried out by a special program called a compiler.

The huge progress in computational chemistry in the last few decades has relied on a number of factors, including the development of new theories and efficient programs to work out these theories. One highly important factor, though, is the huge growth in the power of computers, as measured by the clock speed of the CPU and the capacity of the memory and of the storage. Actually, computational chemistry started out *before* electronic computers were available. Douglas Hartree (1897–1958), a British mathematician whose work will be discussed in Chapter 2, built a 'computer' based on cogs and wheels whose motion—based on mechanical principles—could be used to work out calculations more rapidly than by hand. This type of computer was already in use in the 1930s, by Hartree and others, including a British physicist, Bertha Swiles (1903–1999), to calculate the distribution of electrons in atoms based on quantum mechanics.

As well as calculations of the quantum mechanical properties of electrons within atoms, computers were used from early days to model potential energy surfaces for interactions between atoms, both to describe simple elementary reactions (such as atom exchange in $H + H_2$, $H_A + H_B - H_C \rightarrow H_A - H_B + H_C$) and to model the behaviour of collections of atoms such as liquids. Obviously, for all these applications, much more complicated calculations could be carried out once electronic computers became available after the Second World War. Since then, computers have undergone huge progress, with the speed of their CPUs and the amount of storage and memory all increasingly roughly exponentially in time following Moore's Law. This tremendous progress has largely been a result of being able to make computer chips containing more and more transistors—each of which has become smaller and smaller. Recently, clock speeds have stopped increasing, but computers continue to become faster, in part due to parallelization, whereby multiple CPUs and multiple cores on each CPU are programmed to carry out multiple operations at the same time.

As well as the huge progress in computer power, there have also been a lot of improvements in the type of theoretical model put forward for exploration by computers, in the algorithms designed to work out the predictions of the models, and in the programs that implement these algorithms. An efficient program needs to take into account the available cores and CPUs, the available memory and storage for values generated during the calculation, and the speed with which data can be exchanged between them. Given a particular theory, there can be many ways to implement it in a program. If this is done in a suboptimal way, the resulting program may be too slow, or may require an unreasonable amount of memory in order to be able to execute, or may need too much storage. Together, the developments in computers, theories, and programs have enabled computational chemistry to take up a central place in chemistry research and education.

Theory and computation play different roles in chemistry, but their points of focus do overlap considerably, to the extent that some people think that theoretical and computational chemistry are the same thing. The present book will focus on the computational aspects, rather than on the theoretical ones. This means that only a minimal amount of basic *theory* of chemistry will be covered—material covered in standard textbooks will by and large be assumed to be known and understood, though it will usually be summarized where needed for convenience and clarity. Mathematical equations will be used where needed to define the background theory or the computational techniques used to work it out. However, in general, a verbal description of the problem will also be included, so that the equations can to some extent be ignored. Also, results from actual calculations will be presented and assessed in the text, with enough details given to enable interested readers to repeat and check the work.

The choice to focus on *computation* rather than *theory* means that the book will not in itself equip the reader to carry out development of new computational techniques, since doing so usually requires extensive knowledge of the theoretical underpinning of a particular area, which is the topic of many other books. What the book does aim to achieve is that readers gain confidence in using computational methods, by understanding how they build on theoretical principles and are used to understand chemistry.

1.4 Different types of computational method

There are many situations in chemistry where computers are used in the ways discussed in the previous sections. The most common ones all involve modelling the behaviour of chemical systems that have many parts, which interact in complicated ways. Molecules (or assemblies of molecules) are made up of nuclei and electrons. The motions of the electrons within the molecules are the subject of *quantum chemical* calculations, which will be described in Chapters 2 and 3. Atoms themselves can also move, as a result of thermal motions or chemical reactions. Chapter 6 describes how the laws of physics can be used to describe these motions, which in turn

Moore's Law This 'law' is actually an empirical observation, whereby the density of components in state of the art computer chips roughly doubles every two years.

requires an understanding of the function giving the energy of the system as a function of the atomic positions—the potential energy surface, which is described in Chapter 5. Motions leading to *chemical reactions* are particularly important, and particular computational methods can be used to describe them—see Chapter 7. To describe the energetics that determine atomic motion, and the structure of molecules, it is possible to use quantum chemical methods, but it is also possible to use simpler *molecular mechanical* methods—and these are the subject of Chapter 4. There are also *hybrid methods* that combine elements of quantum mechanical and molecular mechanical approaches, as described in Chapter 8.

This short book will not cover every aspect of computational chemistry. For each of the areas mentioned above, many specialized techniques will not even be mentioned—including, no doubt, many topics that would be considered to be essential by others. Also, some uses of computers in chemistry will not be discussed at all. Foremost amongst these is probably the use of computers to manage, access, and learn from large databases of information concerning chemistry. Used for anything from predicting nuclear magnetic resonance spectra for a new compound, to designing new inhibitors of an enzyme as a drug candidate, to modelling the likely three-dimensional structure of a biomolecule, this type of method is widely used nowadays in chemistry as in other fields.

1.5 A note on mathematics

Computational methods are based on fundamental physical theories of quantum and classical mechanics. These theories as well as the approximate computational methods are most precisely expressed in terms of mathematical equations, and indeed most books on theoretical chemistry contain a very large number of such equations. Writing computer programs to implement theoretical methods requires insight into the corresponding physical models as well as an in-depth understanding of the associated mathematics. On the other hand, to *use* existing programs, only a qualitative physical understanding of what a particular method or approximation entails is mandatory. The aim of this book is to provide this understanding for the key computational approaches, while keeping the mathematical description as light as possible. Still, the book does involve a significant number of equations, and the meaning of all symbols used in these equations is given in the text. Occasionally, the equations are written in a slightly simplified way that makes their key message clearer while somewhat sacrificing mathematical rigour. Interested readers are encouraged to consult the suggested 'Further reading' materials.

1.6 Exercises and test calculations

As well as learning about the theory behind computational chemistry, an improved understanding of the field also requires running calculations oneself. To support this activity, this book contains detailed descriptions of many simple calculations, performed by the author while writing this book, in order to provide some insight into different types of calculations. Most of these sample calculations were chosen to be quite simple—well below the 'state of the art' of calculations performed in research projects at the time of writing. With increasing computer power, such calculations will presumably become even easier to perform. This is not a disadvantage: to *learn* about computational methods, the best calculations are those that execute very rapidly, within minutes or less, so that the user rapidly gains expertise in understanding the link between the chosen calculation parameters and the output results. Running more demanding calculations is usually no more conceptually challenging. Almost all of the sample calculations were performed using free or widely available software, and sample input files for many of the calculations are available online.

The book also contains a number of *exercises* at the end of each chapter, in the form of rather open-ended 'mini-projects'. As just mentioned, many programs are available for carrying out the various types of calculation described in this book. Some programs can be obtained completely freely (though some computing expertise may be needed in order to compile them or otherwise make them ready for use). University chemistry students will probably also have access to computers on which commercial software for performing calculations is installed. While the choice of software can be important in research projects due to differences in efficiency or in method availability, for the purposes of *learning*, the particular software package used is of less importance. Because software codes continue to develop and because each university will have different software available, none of the instructions for the exercises in this book are specifically tailored to a particular program. 'Tutorials' for different programs can however readily be found in the documentation of the programs or elsewhere on the internet, and can be used in combination with the book.

1.7 Further reading

There are a number of textbooks covering all core aspects of computational chemistry in a more detailed way than in the present short book. Most of the content of this book is present in more extensive form in the following two references, which can therefore be used in parallel with every chapter of the book:

- *Introduction to Computational Chemistry*, 3rd edition, Frank Jensen, Wiley, 2017.
- *Essentials of Computational Chemistry: Theories and Models*, 2nd edition, Christopher J. Cramer, Wiley, 2004.

The following book gives a general introduction to the background theoretical concepts used in computational chemistry:

- *An Introduction to Theoretical Chemistry*, Jack Simons, Cambridge University Press, 2003.

1.8 Exercises

1.1 Use a computer together with a plotting program (a spreadsheet program can be well suited for this) to evaluate equation (1.2) for different values of r. Use $k = 800$ kJ mol^{-1} Å^{-2}, $r_0 = 1.5$ Å. For any given r, such as $r = 2$ Å, V can be evaluated very readily without a computer, but to evaluate it for many values of r between 0.5 and 2.5 Å, a computer approach will be preferable— though in this simple example, it is clear by inspection that a parabola will be obtained, so a sketch could be generated by hand, by taking a few values. For a more complicated expression, such manual analysis will be harder. Consider the more realistic Morse curve (equation 1.3), and generate a plot using $r_0 = 1.5$ Å, $D = 400$ kJ mol^{-1}, and $\alpha = 1$ Å^{-1}, and considering values of r between 0.5 and 6.0 Å:

$$V = D\left\{1 - \exp\left(-\alpha\left(r - r_0\right)\right)\right\}^2 \qquad (1.3)$$

1.2   In section 1.2, it was explained that problems for which computers are needed are those where too many computations need to be carried out in order to test a particular theory. A very simple example of this can be given for the problem of finding all the *prime* numbers that are smaller than some threshold N. For small N, the answer can be found by inspection, while for intermediate values of N, no computer may be needed, but paper and pencil and some patience will be. For large N, a computational approach will become mandatory. Try to write a simple program method to identify all the prime numbers up to $N = 1,000,000$.

1.9   Summary

- To work out the predictions of many useful chemical theories, a large amount of computation is needed.
- This is particularly true for theories seeking to predict experimental behaviour based on the quantum mechanical properties of molecules as assemblies of atomic nuclei and electrons.
- Theories based on the properties of atoms and molecules and their interactions typically also require extensive computation.
- This book focuses on describing many of the popular computational techniques in chemistry.

2 Quantum Chemistry

2.1 Introduction

Pretty much all of computational chemistry relies on quantum mechanics in the sense that molecular systems follow the laws of quantum mechanics. But 'quantum chemistry' has a more specific meaning: it is the study of chemistry through the use of approximate solutions to the electronic Schrödinger equation. The topic of this chapter is to describe the background and principles of quantum chemistry, with a focus on the conceptually most important approximate approach, Hartree–Fock theory. Other quantum mechanical techniques will be introduced and assessed in Chapter 3.

In the most general way, the time-dependent Schrödinger equation (2.1) describes how to predict the evolution in time of a system based on its *wavefunction* Φ. The wavefunction relates to the positions of the particles in the sense that the square of the wavefunction measures the probability that the N particles in the system adopt a set of positions given by the vector of coordinates $R = (R_{1,x}, R_{1,y}, R_{1,z}, R_{2,x}, \dots R_{N,z})$. The Hamiltonian operator, $\hat{H}$, contains a set of terms relating to kinetic and potential energy that, when acting on the wavefunction, describe the way in which this wavefunction should vary in time. Hence the right-hand side of equation (2.1), which can be calculated if one knows the wavefunction, tells us the way in which the wavefunction is changing in time.

$$i\hbar \frac{\partial \Phi(R,t)}{\partial t} = \hat{H}\Phi(R,t) \tag{2.1}$$

In principle, the time-dependent Schrödinger equation could be used to predict what happens in chemical reactions. Take as an example electrophilic addition of a methyl cation CH_3^+ to a molecule of benzene C_6H_6. This is a system with 66 particles, namely 16 nuclei and 50 electrons (six for each carbon atom, one for each hydrogen atom, minus one for the positive charge). So the vector of coordinates has 198 elements, one each for the x, y, and z coordinates for each particle. Imagine that you know the wavefunction $\Phi(R,t_0)$ corresponding to the initial conditions. The square of this wavefunction would have large values for positions of electrons and nuclei that would occur in the separated reactants. To predict what happens next, the right-hand side of the equation could be evaluated, by applying the Hamiltonian operator to the wavefunction. This would yield a predicted rate of change of the wavefunction (its time derivative). In turn, this could be used to update the wavefunction to a value a short time later, $\Phi(R,t_0 + \delta t)$, and this procedure could be repeated many times in order to predict the whole evolution of the system.

In practice, this approach is not practical and is almost never used. Instead, two modifications are made that lead to a more manageable problem. First, because for many systems time-*invariant* wavefunctions provide a reasonable description, the time variable is removed, leading to the time-independent Schrödinger equation. Next, the wavefunction is separated into two parts. One wavefunction $\Psi(r)$ describes the distribution of the *electrons* around the nuclei, for a given set of coordinates R_N of the nuclei. This wavefunction only depends explicitly on all of the coordinates of the electrons (denoted here collectively as 'r'), though the function is different for each nuclear arrangement. A different wavefunction $\Theta(R_N)$ describes the distribution of the nuclei. This is the Born-Oppenheimer approximation, whereby the overall wavefunction can be written as a product of $\Psi(r)$ and $\Theta(R_N)$. Separate Schrödinger equations can be written for the two wavefunctions. In this chapter, we will be concerned with the *electronic* Schrödinger equation:

$$\hat{H}_{elec}\Psi(r) = E\Psi(r)$$

(2.2)

Solving equation (2.2) for a given set of values for the coordinates R_N of the nuclei yields a wavefunction $\Psi(r)$ describing the probability of different arrangements of the electrons, and the energy E associated with the electrons. Taking a simple example again, solving equation (2.2) for a molecule of lithium hydride involves assuming positions for the Li^{3+} and H^+ nuclei, then finding the function $\Psi(r) = \Psi(x_1,y_1,z_1,x_2, ... z_4)$ that depends on the 12 coordinates of the four electrons, which is such that when acting on it with the electronic Hamiltonian operator, the same function is obtained, multiplied by a constant, which is then the energy of the system. Equation (2.2) always has multiple different solutions, with different wavefunctions and different energies. For typical molecules, the solution with the lowest energy—the so-called *ground state*—is characterized by a much lower energy than that of any of the other solutions. Methods to find the ground state wavefunction and energy will be the main focus in this chapter.

The apparent simplicity of equation (2.2) disguises its highly complex nature. The electronic wavefunction of LiH is a function of twelve variables, while for the benzene + methyl cation example, the wavefunction depends on the coordinates of 50 electrons, so it is a function with 150 variables. The corresponding Hamiltonian operator contains something like 2,000 terms—to apply it to the wavefunction, one needs to do things like taking the second derivative with respect to the x, y, and z coordinates of each of the electrons, to divide the wavefunction at each value of the electronic coordinates by the distance between each pair of electrons, and so on. Only for the very simplest of problems can this equation be tackled using paper and pencil, and even then, one can generally only obtain approximate solutions. For the molecules that chemists are interested in, one needs to use approximate methods and computers to generate the solutions.

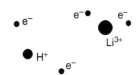

2.2 Hartree–Fock theory

There are many approximate techniques to solve the electronic Schrödinger equation. Almost all of them are linked in some way to one particular approximate approach, the Hartree–Fock (HF) method, and we will therefore focus on this method in this chapter.

The basic approximation in Hartree–Fock theory (and its predecessor, Hartree theory) is an assumption that electrons move independently of one another throughout the molecular system. This may appear to be a very strange choice of approximation, since Coulomb's Law means that electrons interact very strongly (and repulsively) with each other. However, electrons move *very fast* within molecules, so that, despite their strong interactions, the independent motion approximation is quite a good one.

The approximation is sometimes also called the *molecular orbital* approximation. This is because assuming independent motion is equivalent to assuming that the over-all wavefunction Ψ for all the n electrons can be written as a *product* (2.3) of simpler wavefunctions ψ for individual electrons, called molecular orbitals:

$$\Psi(r_1, r_2, \ldots, r_n) \equiv \psi_1(r_1) \times \psi_2(r_2) \times \ldots \times \psi_n(r_n) \tag{2.3}$$

This introduces a significant simplification of the wavefunction, which also makes the problem of determining its shape much more tractable. One does now need to determine the shape of many molecular orbitals—as many as there are electrons—but each of them is far simpler than the overall wavefunction. The individual orbitals are chosen to be *orthogonal* to one another (any integral over all space of the *product* of two different orbitals must be equal to zero).

Equation (2.3) is not quite correct in two ways that are linked to the fact that in quantum mechanics, electrons behave as though they are *spinning*. The speed at which they spin is quantized, and this needs to be taken into account by introducing both a spin coordinate ω and a spin part to the wavefunction, so that molecular orbitals $\psi(r)$ need to be replaced by molecular spin orbitals $\chi(r,\omega)$. These can be written as a product of a spatial part (simply $\psi(r)$) and a spin part, which can be written as $\sigma(\omega)$, or as either $\alpha(\omega)$ or $\beta(\omega)$, corresponding to spin 'up' or 'down'; the 'ω' is often left out so one can also simply write α or β. The combined spatial and spin variables for a given electron, (r_i, ω_i), are often denoted 'x_i'. Also, all electrons need to be treated as being very strictly identical to each other, so the overall wavefunction cannot treat individual electrons differently. This requirement can be accommodated through a slight modification to the above product of orbitals expression, whereby the approximate wavefunction is written as a combination of all possible such products, with the electrons permuted between the orbitals. The resulting combination can be chosen so that it is *antisymmetric* with respect to swapping the coordinates of any two of the electrons. For a four-electron system, there are six pairs of electrons, so one needs to ensure that the wavefunction follows six relations such as:

$$\Psi(x_1, x_2, x_3, x_4) = -\Psi(x_1, x_3, x_2, x_4) \tag{2.4}$$

For a system of n electrons, the combination of product terms that satisfies this condition upon swapping *any* two electrons can be written as the *determinant* of a matrix, with n rows and columns. This is shown in equation (2.5). The use of determinants was introduced by the American physicist John Slater, so wavefunctions of the form of equation (2.5) as used in Hartree–Fock theory are often referred to as Slater determinants.

$$\Psi(x_1, x_2, \ldots x_n) = \frac{1}{\sqrt{n!}} \begin{vmatrix} \chi_1(x_1) & \chi_2(x_1) & \cdots & \chi_n(x_1) \\ \chi_1(x_2) & \chi_2(x_2) & \cdots & \chi_n(x_2) \\ \vdots & \vdots & \ddots & \vdots \\ \chi_1(x_n) & \chi_2(x_n) & \cdots & \chi_n(x_n) \end{vmatrix} \tag{2.5}$$

The requirement that the wavefunction be antisymmetric is a consequence of the Pauli principle. It has a number of quite far-reaching effects, one of which is that in a given molecular system, no more than two electrons may be described by the same molecular orbital ψ. Despite the significant difference between equations (2.3) and (2.5), though, the underlying formalism whereby the wavefunction is a product of one-electron molecular orbitals still determines many of the properties of Hartree–Fock theory, and in this chapter we will mostly consider the Hartree–Fock wavefunction to be a simple product, even though in reality it is a Slater determinant.

The Hartree–Fock approximation is a valuable one to make, because it makes it possible to 'solve' the electronic Schrödinger equation. The price to pay for this is that you do not obtain an exact solution, because the true wavefunction for a many-electron system is not a Hartree–Fock wavefunction. What, then, do you get? To understand this, it is useful to consider an expression for the energy. For an *exact* solution $\Psi(r)$ to the Schrödinger equation, the energy E obtained by solving equation (2.2) can be rewritten according to equation (2.6):

$$E = E \times \int \Psi^2(r)\,dr = \int \Psi(r)E\Psi(r)\,dr = \int \Psi(r)\hat{H}_{elec}\Psi(r)\,dr \tag{2.6}$$

In this expression, the first equality is due to the fact that the wavefunction is *normalized* (the sum of probabilities associated with all possible combinations of coordinates, or the integral over all possible values of r of the square of the wavefunction, is equal to one). For the second equality, because E is a constant, it has simply been inserted into the integral, and the square has been written out explicitly (note that in general the square of the wavefunction should be replaced by the square of its magnitude, because in some cases the wavefunction can adopt complex values, but this aspect is neglected in equation (2.6) and almost everywhere else in this book). For the third equality, the fact that Ψ is a solution of the Schrödinger equation, so that $E\Psi = \hat{H}\Psi$, has been used.

The last step above is not applicable for a function that is not an exact solution of the Schrödinger equation. Instead, one can *define* the 'energy' E_{approx} corresponding to a given approximate wavefunction Ψ_{approx} using the same sort of expression:

$$E_{approx} = \int \Psi_{approx}(r)\hat{H}_{elec}\Psi_{approx}(r)\,dr \tag{2.7}$$

It can be shown that this approximate energy must necessarily be *higher* than the true ground-state energy of the system. This is the *variational principle*, and it is central to Hartree–Fock theory. The Hartree–Fock wavefunction of a given system is defined as the Slater determinant composed of the set of orthogonal molecular orbitals that return the lowest possible energy. Though this energy will be larger than the true ground-state energy of the system, it will be the energy that is closest to the real one that can be obtained for a wavefunction of the form of equation (2.5).

Inserting the Slater determinant expression of equation (2.5) for a system with n electrons into equation (2.7) leads, after much algebra, to the following expression for the energy:

$$E = \sum_{i=1}^{n} h_{ii} + \sum_{i=1}^{n}\sum_{j=i+1}^{n} (J_{ij} - K_{ij}) \tag{2.8}$$

In this expression, the symbols h_{ii}, J_{ij}, and K_{ij} refer to various integrals carried out over the different molecular orbitals. Each of the integrals h_{ii} only depends on the shape of a single molecular orbital χ_i or ψ_i, and these integrals are called 'one-electron' terms. They provide a measure of the energy due to coulombic

Atomic units

In equation 2.9, to avoid overloading the equation, *atomic units* have been assumed. This is a set of units that is often used for convenience in theoretical chemistry instead of the more usual SI units. Equation (2.9) relates to Coulomb's Law for the interaction between two electrons, and instead of the term $1/|r_1 - r_2|$, you might expect instead to find $1/4\pi\varepsilon_0 \times q_e^2/|r_1 - r_2|$, where ε_0 is the dielectric permittivity of vacuum, and q_e is the charge of the electron. The atomic unit for permittivity is such that the numerical value of $4\pi\varepsilon_0$ is 1, and the atomic unit of charge is such that the numerical value of q_e is −1. In principle, to satisfy the dimensionality of the equations, it would be more correct to still include these symbols in equations, but leaving out symbols whose value is equal to one leads to such a simplification that it is an almost universal practice in quantum chemistry. Two of the atomic units are especially important when running computations, as input and output from programs is often expressed in these units. The atomic unit of energy is called the hartree (named after Douglas Hartree), and is equal to 4.36×10^{-18} J, or 2625.5 kJ mol^{-1}. The atomic unit of length is called the bohr (it is the distance between the nucleus and the electron in the model proposed by Niels Bohr for the ground state of the hydrogen atom), is sometimes written a_0, and is equal to 0.529 177 Å.

interaction between an electron occupying that orbital and the positively charged nuclei in the system, summed with the kinetic energy of the electron. The integrals J_{ij} require consideration of two molecular orbitals each, and provide a measure of the Coulombic repulsion energy between the electrons occupying these two orbitals:

$$J_{ij} = \iint \psi_i(r_1)\psi_j(r_2)\frac{1}{|r_1-r_2|}\psi_i(r_1)\psi_j(r_2)\,dr_1dr_2 = \langle \psi_i\psi_j | r_{12}^{-1} | \psi_i\psi_j \rangle \qquad (2.9)$$

In this expression, $|r_1 - r_2|$ is the distance between points r_1 and r_2, also written as r_{12}. Integrals such as equation (2.9) arise in many contexts, so various forms of shorthand are used to write them—here one such notation was included on the right-hand side. The integrals K_{ij} also depend on two orbitals, and have an expression similar to that for the integrals J_{ij}, but they are numerically much smaller in magnitude. They provide a correction to the J_{ij} Coulombic repulsion between electrons that arises due to the antisymmetrization of the wavefunction.

2.3 The Hartree–Fock method in practice

There are many ways to describe the procedure (or algorithm) used to find the Hartree–Fock solution for a given molecular system. Here, to give a better understanding of this important procedure, we will use three different levels of description. The **first and most general** one is the one given above: a starting set of orbitals are chosen using some or other procedure (perhaps simply a guess based on chemical intuition for the bonding in the target system), a Slater determinant (equation 2.5) is constructed, its energy is evaluated using equation (2.7) (or more precisely (2.8)), then changes are made to the orbitals (while keeping them orthogonal to one another), and this is repeated until the lowest possible energy has been reached. In practice, though, this would require an unreasonable number of steps, and defining how to vary the orbitals is not trivial.

The **second level of description** is more detailed, and involves the *Fock equations* (2.10). These are obtained by inserting the expression for the approximate form of wavefunction used (equation 2.5) into the electronic Schrödinger equation and rearranging (in a way which is not described in detail here).

$$\hat{f}\psi(r) = \varepsilon\psi(r) \qquad (2.10)$$

In this expression, $\hat{f}$ is the Fock operator (more precisely, the one-electron Fock operator), a modified form of the Hamiltonian operator, ε is an energy, and $\psi(r)$ is a molecular orbital. Equation (2.10) is an equation whose solutions ε and ψ are one-electron properties, and in this it appears very different from the many-electron Schrödinger equation. However, the orbitals obtained by solving the Fock equations are those that yield the Hartree–Fock many-electron wavefunction. This is because the Fock operator contains within it terms that describe the electron–electron interactions.

To solve equation (2.10) efficiently using a computer, the molecular orbitals need to be expressed as combinations of simpler functions, the *basis functions* $\phi(r)$:

$$\psi_i(r) = \sum_{j=1}^{n_{basis}} c_{ij}\phi_j(r) \qquad (2.11)$$

The n_{basis} functions $\phi(r)$ are typically chosen to resemble each of the occupied atomic orbitals of the atoms making up the system. This collection of functions is called the *basis set*, and the way in which such basis sets are created and chosen in calculations is described below, in section 2.6. The important point here is that the functions $\phi(r)$ are fixed: they are chosen before the calculation is carried out, and their shape does not change at all during a calculation.

What changes is the set of coefficients c_{ij} which determine the extent to which a particular basis function ϕ_j contributes to a particular molecular orbital ψ_i, and these coefficients (together with the energies ε) are now the unknowns that are being solved for. Hence the problem of searching for a set of functions ψ has been replaced by the problem of searching for a set of *numbers* c_{ij}. Computers are not very good at handling variations between many different functions, but manipulating collections of numbers is straightforward. Indeed, by expressing the molecular orbitals as *linear* combinations of atomic orbitals, there turns out to be a very simple way to implement the variational principle. Inserting equation (2.11) into equation (2.10) yields a set of linear equations which can be solved by carrying out appropriate operations on a set of numbers collected in various matrices, to yield the lowest energy solution according to the variational principle.

There is however a twist: as mentioned above, the Fock operator contains within it a description of the electron–electron interactions. As a consequence, the Fock operator depends on the shape of the orbitals, i.e. on the coefficients c_{ij} which are themselves the unknowns in the Fock equations! This might make it seem impossible to write down let alone solve these equations. However, on closer inspection, the equations *can* be solved, by using a set of iterations. First, the values of the coefficients c_{ij} are guessed, yielding an initial set of orbitals. These are used to build the Fock operator, and the Fock equations are solved to yield a new set of orbitals. These are then used to build a new Fock operator, which will typically be somewhat different from the first one. The new set of Fock equations are then solved, and this cycle is repeated until the orbitals no longer change significantly from one step to the next one. At this point, the Hartree–Fock equations have been solved.

In this second description, the iterations are not used to implement the variational principle, as in the naïve first description given above. At each cycle, solving the Fock equations leads to obtaining the fully variational solution. Instead, iterations are needed so that the molecular orbitals used to construct the Fock operator become consistent with the ones obtained by solving it. This aspect explains why the Hartree–Fock method is sometimes called the **self-consistent field (SCF)** method. Also, the use of molecular orbitals expressed as linear combinations of atomic orbitals as in equation (2.11) justifies the use of the acronym LCAO-MO (linear combination of atomic orbitals-molecular orbitals) to refer to Hartree–Fock calculations.

There is a **third, more practical, level of description** of the Hartree–Fock method, which focuses on techniques designed to improve the reliability and speed of calculations. There are many aspects here that a user will need to be acquainted with in order to run calculations effectively. A first important factor concerns the **initial guess** for the orbitals. The closer they are in shape to the final Hartree–Fock orbitals, the more likely the procedure is to converge and to converge rapidly. A common choice of initial guess is the converged orbitals from a previous calculation on a similar system. For example, the orbitals obtained when using a smaller set of basis functions $\phi(r)$ can be used. Another option is to use the orbitals obtained by solving the Fock equations (2.10) but with a simplified Fock operator (e.g. based on methods such as Hückel theory) which does not depend on the orbitals, so that no iterations and no initial guess are needed for it.

A second aspect concerns the convergence of the iterative solution of equation (2.10). In favourable cases, the repeated solution of equation (2.10) leads to smaller and smaller changes in the orbitals obtained and hence in the energy. However, it can happen that even after several cycles of application of equation (2.10), significant changes in the orbital shapes and in the energy are still occurring, and a consistent solution is only reached after a large number of cycles, increasing the computational effort. The iterations can even behave chaotically, with large changes in the orbitals occurring from the start or after some iterations.

Electronic structure programs use several numerical tricks to avoid these problems. In one type of approach, information derived from several sets of orbitals $\psi^{(i-1)}$, $\psi^{(i-2)}$, and so on, and not only from the latest set $\psi^{(i)}$, is used in order to construct the Fock operator for the next iteration. A typical method of this type is known as the direct inversion in iterative subspace (**DIIS**) method. In another commonly used approach, the orbital energies $\varepsilon^{(i)}$ obtained in cycle i are slightly modified before constructing the Fock operator for the next cycle, by increasing the energy for all empty orbitals by a fixed amount. This is called **level shifting**. A combination of methods, such as DIIS and level shifting, as well as careful selection of the initial guess will usually lead to rapid and robust convergence, but trial and error of several procedures may be needed for problematic species such as those containing some transition metals.

Another aspect to do with computational efficiency of the Hartree–Fock method concerns the evaluation of the Fock operator in equation (2.10), and of the matrix of values arising from applying this operator to the various orbitals (the Fock matrix). In order to construct each of the elements of the Fock matrix, integrals over the molecular orbitals must be used, see equations (2.8) and (2.9). These integrals can be evaluated, manipulated, and stored in various different ways. Usually what happens is that the integrals of equations (2.8) and (2.9) are obtained from combining values of integrals over the *basis*

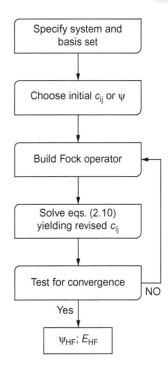

Specify system and basis set

Choose initial c_{ij} or ψ

Build Fock operator

Solve eqs. (2.10) yielding revised c_{ij}

Test for convergence

NO

Yes

ψ_{HF}; E_{HF}

functions. Indeed, inserting the expression of equation (2.11) into equations (2.8) or (2.9) leads to sums of such integrals. Equation (2.8) yields the so-called *one-electron integrals*:

$$\int \phi_a(r)\hat{h}\phi_b(r)dr \qquad (2.12)$$

In this expression, $\hat{h}$ is a one-electron Hamiltonian operator, and ϕ_a and ϕ_b are two basis functions. If there are n_{basis} basis functions, then there are roughly $n_{basis}^2/2$ one-electron integrals (the factor of two arises because switching ϕ_a and ϕ_b does not change the value of the integral, so only one of the two combinations needs to be evaluated). Each of the integrals given by equation (2.12) can contribute to several different elements of the Fock matrix. This makes it desirable for these numbers to be computed and stored in memory, so as to be used whenever needed. For typical values of n_{basis} of the order of a few hundred, this is not a problem with modern computers, since there will be fewer (or much fewer) than a million such numbers—which can hence be stored in less than a few megabytes of memory. Equation (2.9) yields the *two-electron integrals*:

$$\iint \phi_a(r_1)\phi_b(r_2)\frac{1}{|r_1-r_2|}\phi_c(r_1)\phi_d(r_2)dr_1dr_2 = \langle \phi_a\phi_b \,|\, r_{12}^{-1} \,|\, \phi_c\phi_d \rangle \qquad (2.13)$$

As in equation (2.9), one of the many forms of shorthand used to denote such integrals is shown on the right-hand side. The expression of equation (2.9) depends on *two* molecular orbitals, each of which occurs twice in the integral. The two-electron integrals of equation (2.13) that arise from inserting equation (2.11) into equation (2.9) can depend on up to *four* basis functions, ϕ_a, ϕ_b, ϕ_c, and ϕ_d. This means that there are roughly $n_{basis}^4/8$ two-electron integrals (the factor of eight is due to the fact that some permutations of a, b, c, and d correspond to the same integral). As for the one-electron integrals, each of the two-electron integrals can contribute to several different elements of the Fock matrix, so an ideal situation would be to be able to calculate and store all of the integrals, then recall them from memory or storage where needed. However, the fourth power dependence on the number of basis functions means that this is not always possible. For $n_{basis}=200$, there are already roughly 2×10^8 two-electron integrals, requiring over a gigabyte of memory, and this climbs very steeply for larger numbers of basis functions.

When the two-electron integrals are stored on a hard disk or other type of long-term storage, this can lead to an inefficient code, due to the (relatively) slow rate of data exchange between the disk and the CPU. For such cases, it can be more efficient to *recalculate* the two-electron integrals as and when they are needed, removing the need to access the slow storage (and removing the need for storage altogether in cases where the number of integrals of equation (2.13) is such that it would exceed available storage). This can be done in combination with a pre-screening algorithm. Based on the position in space where the basis functions ϕ_a, ϕ_b, ϕ_c, and ϕ_d are centred, the nature of these functions, and the expression for the molecular orbitals, it is possible to evaluate whether the resulting two-electron integral will make a contribution to the energy that is smaller in magnitude than some threshold. This pre-evaluation or screening of the importance of the integral can be done in a way that is much less computationally demanding than actually evaluating the integral.

In so-called *direct* methods, most of the two-electron integrals are not stored and re-used at each cycle, but instead at each SCF cycle the contribution of each integral is estimated in this way. If it is considered to make a negligible contribution it is omitted, while otherwise it is calculated. Especially for large molecules, the screening leads to a sharp reduction in the number of integrals that actually need to be evaluated. Overall, even though some of the two-electron integrals need to be recalculated many times, direct approaches can yield faster program execution and they certainly require much less storage. Many other advanced techniques, beyond the scope of this book, are available to minimize the computer time needed to carry out a Hartree–Fock calculation, and the user will come across many corresponding options that need to be specified in input files.

An example Hartree–Fock calculation: cholesterol

To illustrate some of the preceding information, consider a Hartree–Fock calculation on a molecule of cholesterol, $C_{27}H_{46}O$ (see ball-and-stick model below). This system contains 216 electrons, and a set of 344 basis functions ϕ are used in order to expand the molecular orbitals according to equation (2.11) (this is the 6-31G basis mentioned in the box 'Some common basis set families' under section 2.6). The main input needed for the program is the coordinates of each of the 74 atoms. A graphical program can be used to draw the molecule and generate a guess for these coordinates. The user also specifies the basis set, and the number of spin-up and spin-down electrons (here each equal to half the total number of electrons).

In the calculation performed here, all two-electron integrals of equation (2.13) were evaluated, and this required storage of a 5 GB file on disk. The Fock equations (2.10) were solved iteratively, requiring 12 iterations—this was enough to converge the energy to within better than 10^{-9} hartree, and to obtain also consistent orbitals with a very low convergence threshold. The calculation required about two minutes using a simple laptop computer. Repeating the described 'direct' procedure drastically reduced the amount of disk storage needed to a few MB. However, it also made the calculation slower, requiring roughly 4 minutes. The relative amount of time needed for the 'direct' and standard procedures varies depending on the program, the molecule, the basis set, and the computer, though for large molecules or basis sets, only the 'direct' approach may be possible since the standard procedure may require too much disk space.

Note that the program automatically used an extrapolation procedure to speed up convergence, leading to converged orbitals for cholesterol within 12 cycles. By turning off all such procedures, more than double the number of iterations were needed, i.e. 29. Still, given that cholesterol is a molecule with a simple bonding pattern and no unpaired electrons, convergence could be achieved simply by repeated application of equation (2.10) without the need to use convergence acceleration procedures. The $C_{21}H_{15}FeN_4OS$ model for the haem-based active species of the cytochrome P450 enzymes shown here is an example of a more challenging species, as it contains a transition metal and three unpaired electrons. Hartree–Fock calculations on this 235-electron system using

(continued...)

a basis set with 310 functions failed to converge at all using the standard acceleration procedures, despite a reasonable initial guess that was fairly close in energy to the final wavefunction. By using level shifting, it was possible to get convergence after 100 SCF cycles.

2.4 The Hartree-Fock wavefunction and energy

Carrying out a Hartree-Fock calculation yields an overall wavefunction (the Hartree-Fock wavefunction or Slater determinant of equation (2.5)), the molecular orbitals (equation 2.11), each with their associated energy ε, and the overall energy (equation 2.7). The orbitals and their energy, as well as the overall energy, are the key outputs that are frequently of interest for understanding the properties of the system.

The overall *energy* of the Hartree-Fock wavefunction is one of the most important outcomes. To appreciate its importance, it is useful to understand what this energy means, and what terms contribute to it. This can be done with reference to equation (2.7). The energy measures the quantum mechanical energy of the electrons making up the system. The electronic Hamiltonian operator is written as a sum of different parts, corresponding to electron kinetic energy, nucleus–electron Coulombic attraction, and electron–electron Coulombic repulsion. The overall energy can also be broken up into these parts, which can be used to analyse the origins of chemical bonding. For chemical purposes, it is important to include also the nucleus–nucleus repulsion energy, which is simply a constant for any given arrangement of the nuclei, but can change a lot between different arrangements, e.g. upon breaking or making a bond. The sum of the energy of the electronic wavefunction and the nuclear repulsion energy is often referred to as the *total energy* in computational chemistry, and equally often, it is called the *electronic energy.*

The reference point on the energy scale, i.e. the point defined as having a total energy of zero, is the one where each of the contributions mentioned above is equal to zero. This is the (hypothetical) situation where all of the nuclei are infinitely far apart (so that the nucleus–nucleus coulombic repulsion energy is zero), all of the electrons are infinitely far from the nuclei, and all electrons are infinitely far from each other. Also, the electrons are stationary, so they have no kinetic energy. This state corresponds to the one obtained upon fully breaking all the chemical bonds in the system, and fully ionizing all the resulting atoms. Breaking bonds and ionizing atoms is very energy-demanding, so the total energy of molecules is much lower than zero. For example, the Hartree-Fock total energy of a water molecule (with O-H bond lengths of 1 Å and a bond angle of 105°, using the 6-31G basis, see section 2.6) is roughly −74 hartree or $-200,000$ kJ mol^{-1}. Unsurprisingly, a large portion of this overall energy is due to the large amounts of energy required to remove the innermost core electrons for the atoms. For the hydrogenic O^{7+} ion, removing the final $1s$ electron requires an energy of roughly 84,000 kJ mol^{-1}. The overall energy of -75.978 hartree calculated for this structure and basis set can be broken down into a nucleus–nucleus repulsion energy of $+8.812$ hartree, an electron–nucleus attractive energy of -198.112 hartree, an electron-electron repulsion energy of $+37.478$ hartree, and an electron kinetic energy of $+75.843$ hartree. These numbers will be slightly different when using a different structure or basis set.

The Hartree–Fock energy is not very accurate in chemical terms, due to the neglect of electron correlation. However, in *absolute* terms, Hartree–Fock produces reasonably accurate results. The error in the total energy is typically somewhat smaller than 1% of the total energy. Because the total energy is huge, this is still a large amount, many thousands of kJ mol^{-1} for most molecules. Fortunately, when comparing energies for the same system with different atomic positions, some of this error cancels out.

Each of the molecular orbitals ψ_i is characterized in terms of its *energy* ε_i and its composition in terms of the basis functions, equation (2.11). The orbital energies are important because they provide an alternative breakdown of the overall energy into contributions from the different orbitals (note that the overall energy of the Hartree-Fock wavefunction is *not* simply the sum of the orbital energies, because additional terms need to be taken into account). Orbital energies are usually negative, with the lowest energies corresponding to molecular orbitals that are largely composed of combinations of core orbitals of the atoms making up the molecule. For example, in the water molecule, the lowest energy orbital energy, with $\varepsilon \approx -21$ hartree, is almost entirely composed of the $1s$ atomic orbital on the oxygen atom. The higher-lying orbitals are valence orbitals, corresponding to bonding and non-bonding (lone pair) electrons. The highest few orbitals, especially the highest occupied molecular orbital or HOMO, are important because molecular properties and reactivity can often be rationalized in terms of the shape of this orbital. Also, the energy of the HOMO provides a rough estimate of the *ionization energy* of the system, IE $\approx -\varepsilon_{HOMO}$ (this is called Koopmans' theorem, after the Dutch mathematician and economist, who suggested it in 1934).

Solving the Fock equations (2.10) yields the *occupied* orbitals that have just been discussed, but it also yields a set of orbitals that are not occupied. These 'vacant' orbitals also provide insight into properties and reactivity, particularly the lowest unoccupied molecular orbital, or LUMO.

The *shape* of the molecular orbitals is often a matter of significant interest. Even though the Hartree-Fock energy is not particularly accurate, the Hartree-Fock orbitals do give a qualitatively reliable description of the electronic structure. Also, they map well onto qualitative models of bonding learned in introductory chemistry classes.

It is helpful to visualize the shape of the molecular orbital functions. However, the functions are not trivial to plot: they depend on all three spatial coordinates, so would require four dimensions in order to be represented in full. As this is not possible, one must instead use some lower-dimensional description. For diatomic molecules, for example, plotting the value of the function along the molecular axis z yields a traditional graph of $\psi(z)$ as a function of z, see Figure 2.1. It is also possible to plot the function on a two-dimensional plane, either by using perspective, or by using contour plots.

Molecular orbital functions are typically large in magnitude near the nuclei in the system being studied. At any given point, they can adopt either positive or negative values. The *relative* sign of the orbital at two points is meaningful and can be understood in terms of the nodes in the atomic orbitals contributing to the molecular orbital, and in terms of the new nodes that arise due to possible anti-bonding character of the molecular orbital. The HOMO of LiH is mainly just the H $1s$ orbital, but it has some σ-bonding character, due to overlap with the Li $2s$ orbital. Also, this orbital needs to be orthogonal to the core Li $1s$ orbital. The main plot (a) in Figure 2.1 shows the large amplitude of the HOMO at the H nucleus, as well as a node close to the Li nucleus that arises due to the required orthogonality with the Li $1s$ orbital. Evidence of bonding overlap with the spatially very extended Li $2s$ orbital is hard to see. The large (but very narrow) negative amplitude near the lithium nucleus misleadingly suggests that this orbital has significant lithium character. Moving away from the internuclear axis by 0.5 (b) or 1.0 Å (c) shows that in fact there is very little lithium character to the orbital.

The *overall* sign for the orbital has no particular meaning, so that the positive and negative regions of the molecular orbital could be swapped without changing the underlying predicted electron density. This is because the electron density contributed by a given molecular orbital is given by the *square* of the orbital function. The electron density falls off very rapidly towards zero as one moves away from the system being studied, and hence so does the magnitude of the molecular orbital functions.

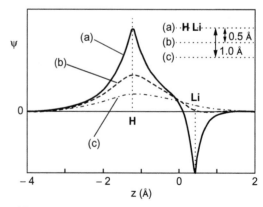

Figure 2.1 Plot of $\psi(z)$ as a function of z for the HOMO of LiH, along three different lines each running parallel to the Li–H internuclear axis. The plot (a) runs through the nuclei, whereas the plots (b) and (c) are displaced by 0.5 and 1.0 Å, respectively.

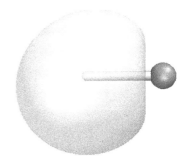

It is also possible to display an orbital by plotting an *isosurface* for the orbital in three dimensions, using perspective to represent the third dimension. An isosurface is the set of points at which the orbital function adopts the same numerical value. The HOMO for LiH is shown in the margin (the isosurface has been made transparent so the H nucleus can be seen). To capture the regions with high electron density, typically two isosurfaces are shown, with the same absolute value, but one positive, and one negative (remember, the positive and negative regions can be swapped without changing anything of the predicted physical properties). The isovalue magnitude, in units of electron$^{1/2}$ volume$^{-1/2}$, is typically chosen so that the isosurfaces enclose the region where the electrons in the orbital being considered spend 'most' of the time, e.g. 90% or so. A typical value is 0.05 electron$^{1/2}$ bohr$^{-3/2}$, which is what is used here for the LiH HOMO. Such isovalue plots yield a reasonably accurate visual impression of where the electrons belonging to an orbital reside. The plot of the HOMO for LiH clearly shows that it mainly has the character of the 1s orbital on H (with a small lobe of opposite sign on Li).

As well as graphical plots, it is also possible simply to look at the coefficients c_{ij} that determine the composition of molecular orbital i in terms of the atomic orbitals ϕ_j, see equation (2.11). With experience, these coefficients can readily be mapped on to isosurface plots. The advantage is that these coefficients are often automatically printed in the output file by a quantum chemical program package, so that no extra step is needed before 'visualizing' the orbitals. Care is needed to take into account the sign conventions for the underlying atomic orbitals. Also, this mapping is easier to perform when using small basis sets of atomic orbitals, as there are then fewer coefficients to consider. Even for larger basis sets, it will usually be the case that only a handful of atomic orbitals make a large contribution to a given molecular orbital. As a rule of thumb, contributions corresponding to coefficients smaller than 0.05 will be so small in terms of electron density that they can be neglected, and indeed, the main contributions will often be larger than 0.2.

Molecular orbitals tend to be spread over the whole of the molecular system— they are somewhat *delocalized*. For molecules which have symmetry, e.g. for which reflection through a plane, or rotation around an axis, leaves their nuclear framework unchanged, the delocalization is imposed by the fact that the orbitals resulting from solving the Fock equations will be symmetrical also: they will either be unchanged by each of the symmetry elements, like the nuclear framework, or their sign will switch.[1] Consider some of the Hartree–Fock orbitals for *p*-dichlorobenzene, shown in Figure 2.2: contributions from *p* orbitals on one of the chlorine atoms are mirrored by ones from the corresponding *p* orbitals on the other chlorine atom. But even for less-symmetric molecules such as camphor, shown in Figure 2.3, orbitals tend to spread over most of the molecule. For example, for a typical organic compound with many C–H bonds, there will be a set of orbitals all apparently composed of different mixtures of the different σ C–H orbitals, rather than having one orbital each for each C–H bond. This can make it hard to analyse bonding directly in terms of orbitals.

As well as the individual orbitals, the *overall* Hartree–Fock wavefunction can also be visualized. This wavefunction is very high-dimensional even for very simple systems. For example, for a hydrogen molecule, with two electrons, the overall wavefunction depends on a total of six coordinates, and a full graphical representation would require seven dimensions. Clearly, this is impossible, so instead, as for molecular orbitals, only

---

[1] For some highly symmetric molecules, the effect of symmetry can be more complicated.

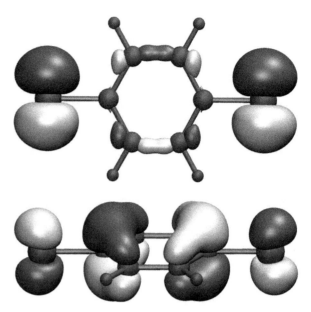

Figure 2.2 Isovalue surface representation of the HOMO-3 (top) and HOMO of *p*-dichlorobenzene, showing the delocalization of the orbital over the whole molecule due to symmetry.

lower-dimensional forms of the wavefunction can be visualized. Lower dimensions can be achieved either by plotting the wavefunction only for certain very specific values of the coordinates, or by first integrating the wavefunction along some coordinates. An example of the first approach will be shown in Chapter 3; here we give an example of the second approach.

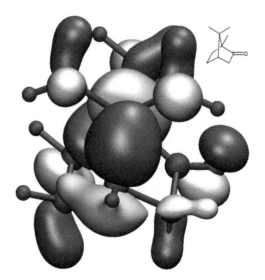

Figure 2.3 Isovalue surface representation of the rather delocalized HOMO-1 of camphor (the structure of camphor is shown as an inset with the same orientation).

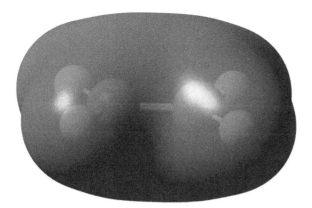

Figure 2.4 Isodensity surface representation of the overall electron density in ethene. The surface shown corresponds to a density of 0.01 e$^-$ bohr^{-3}.

The overall *electron density* $\rho(r)$ is obtained by taking the integral of the square of the wavefunction over the coordinates of all the electrons except one. The electron density for a Hartree–Fock wavefunction is given by the sum of the densities for each molecular orbital, i.e. the sum of the squares of each molecular orbital at each point. This density is a function of the three spatial variables, and can be plotted in various ways, e.g. as in the above orbital plots by showing an isodensity surface, corresponding to all the points where the density adopts a certain value. Such an isodensity plot is shown in Figure 2.4 for the simple case of the ethene molecule, C_2H_4. For the isovalue chosen, you can detect the overall shape of the molecule, including the π electron cloud above and below the molecular plane.

2.5 Restricted and unrestricted Hartree–Fock methods

As discussed above near equation (2.4), the orbitals used to build up the Hartree–Fock wavefunction are *spin-orbitals* $\chi(r,\omega)$, which are products of two functions, a spatial part, $\psi(r)$, and a spin part, simply written as α or β for spin-up and spin-down electrons respectively. Most stable molecules have *closed-shells*: there are equal numbers of spin-up and spin-down electrons, which 'pair up', with an α and a β electron both occupying the same spatial orbital. In this way, the molecule maximizes occupation of the low-energy orbitals (the ones with low ε) while also respecting the Pauli principle. For molecules known to adopt this type of electron structure, this behaviour can be assumed, and used to slightly simplify solving the Hartree–Fock equations. Consider a molecule of ethanol, $CH_3–CH_2–OH$. This has 26 electrons, and is closed-shell. That means that there are only 13 different occupied spatial orbitals for which one needs to find the expansion coefficients of equation (2.11) when solving the Fock equations. Using this simplification to the Hartree–Fock procedure, amounts to carrying out a so-called **restricted Hartree–Fock** calculation.

For systems with an odd number of electrons, or for high-spin systems with two, four, six, or more unpaired electrons (e.g. an oxygen atom in its ground electronic state has two unpaired electrons; it is a spin-triplet), then obviously the electrons will no longer pair up two by two in a given spatial orbital. What then happens in a Hartree–Fock calculation? This is perhaps best understood with an example. Consider

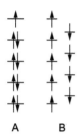

A   B

Figure 2.5 Schematic representation of ROHF (A) and UHF (B) wavefunctions for the methyl radical.

the methyl radical. It has an odd number of electrons, nine, so it must have one unpaired electron. We draw its electronic structure in a Lewis or dot-and-cross diagram as having three valence electron pairs, with one electron remaining unpaired. There is also a pair of electrons for the carbon 1s electrons. There are two main ways to perform a Hartree–Fock calculation on this species. In one limit, one can *assume* that all the electrons except one are paired, and assign the paired electrons two-by-two to the corresponding four spatial orbitals. The ninth electron is assigned to a fifth spatial orbital. Note that the five spatial orbitals (see Figure 2.5A) are required to be different to each other (or *orthogonal*). This type of wavefunction is obtained in a so-called **restricted open-shell Hartree–Fock (ROHF) calculation** (sometimes the 'open-shell' part is omitted, and one needs to infer that the calculation is different than a standard closed-shell calculation from other information).

Another possibility is to simply assume that the wavefunction involves nine occupied molecular spin orbitals, five with α or spin-up character, and four with β or spin-down character. The five spin-up spatial orbitals are required to be orthogonal to each other, and the four spin-down ones also, but no special requirement is placed on the relation between the spatial parts of the α and β orbitals. This approach leads to an **unrestricted Hartree–Fock (UHF) calculation**, and yields *nine* spatial orbitals (Figure 2.5B). In practice, the occupied β orbitals will each be rather similar to one of the α orbitals. However, there will always be slight differences, because the variational principle has been applied to find the lowest possible energy within the allowed range of possibilities for the wavefunction, and allowing α and β orbitals to be different represents an extra degree of freedom that can lower the energy.

One consequence of allowing spin-up and spin-down orbitals to have different spatial parts is that the wavefunction obtained no longer corresponds to a definite or 'pure' spin state. For a system with n_α α electrons and n_β β electrons, then the projection of the overall spin on the z axis, the s_z quantum number, is given by $s_z = \frac{1}{2}(n_\alpha - n_\beta)$. Another way to characterize the spin state of a system is in terms of the quantum mechanical operator for the square of the spin, $\hat{S}^2$. The expectation value of this operator for the wavefunction is equal to $s_z(s_z+1)$ for a pure spin state. For example, for a doublet, with $n_\alpha = n_\beta +1$, the expectation value of $\hat{S}^2$ would be 0.75, or for a triplet, with $n_\alpha = n_\beta +2$, it would be 2. UHF wavefunctions do not correspond to pure spin states, and thereby suffer from so-called *spin contamination*. They return expectation values of the $\hat{S}^2$ operator that are larger than expected based on the number of α and β electrons. The magnitude of the deviation can be understood based on the following equation for the expectation value of $\hat{S}^2$ for a UHF (or ROHF) wavefunction:

$$\int \Psi \hat{S}^2 \Psi \, dr = s_z(s_z + 1) + n_\beta - \sum_{i=1}^{n_\alpha} \sum_{j=1}^{n_\beta} \left(\int \psi_i^\alpha(r) \psi_j^\beta(r) \, dr \right) \tag{2.14}$$

In (2.14), the first term on the right-hand side is the pure spin value of the expectation value of $\hat{S}^2$, the second term is the number of β electrons, and the last term is a sum of the overlap integrals of the spatial parts of all of the occupied α molecular orbitals with all of the occupied β molecular orbitals. These overlap integrals adopt values between 0 and 1: two orbitals that are completely different or *orthogonal* yield an overlap of zero, while two orbitals that are exactly the same have an overlap of one. For an ROHF wavefunction, it is easy to see that the second and third terms cancel out. This is because for each occupied β spatial orbital (i.e. for each doubly

---

Comparing ROHF and UHF calculations

Consider a planar methyl radical with all HCH angles of 120° and C–H distances of 1.1 Å. An ROHF calculation of the energy and orbitals using a particular set of atomic orbitals (the STO-3G basis set, see box titled 'Some common basis set families' under section 2.6) returns an overall energy of –39.07076 hartree. The second-lowest spatial molecular orbital, corresponding to a combination of all three C–H σ bonds, has MO coefficients c_{ij} of 0.652 for the 2s orbital on carbon, and of 0.223 for each of the 1s orbitals on the three hydrogens. The orbital's energy is –0.824 hartree. A UHF calculation for the same system with the same atomic orbitals set and the same structure returns an energy of –39.07552 hartree, lower by about 5 millihartree. The second-lowest spatial orbital for an alpha electron has coefficients of respectively 0.720 and 0.187 for the C 2s and H 1s basis functions, and an energy of –0.897 hartree. The second-lowest β spatial orbital is different in shape and in energy: it has coefficients of 0.581 and 0.260 respectively, and an energy of –0.787 hartree. The expectation value for the $\hat{S}^2$ operator for the UHF wavefunction is 0.767, slightly higher than the value of 0.75 expected for a pure doublet.

---

occupied orbital in ROHF), there is one particular overlap integral, the one with the corresponding α spatial orbital, which is equal to one. All the other overlap integrals are equal to zero. For a UHF wavefunction, this pattern breaks down. For modest spin contamination, it will still be the case that each β spatial orbital will have an α counterpart with which it overlaps quite strongly, leading to an overlap integral of almost—but not quite—one. Hence the second and third terms in equation (2.14) will still almost cancel out, though the third term will be smaller than the second one, so that the expectation value of the $\hat{S}^2$ operator ends up being slightly larger than the pure spin value. Where the α and β spatial orbitals differ more strongly, there can be much larger differences.

When carrying out calculations on open-shell systems, you need to choose between ROHF and UHF methods. ROHF can be preferable as it returns a wavefunction with a well-defined spin. Using ROHF is also more convenient for the purposes of wavefunction analysis, because it is then obvious which orbitals correspond to unpaired electrons, and which do not. However, it is slightly more difficult to write a program to carry out ROHF calculations than it is to write one for UHF calculations, so in many codes, only UHF variants are available for some types of calculations. Also, in many contexts, the lower energy of UHF wavefunctions makes them preferable.

2.6 Basis sets

In order to carry out Hartree–Fock or other quantum chemical calculations, it is useful to consider in more detail the set of basis functions of equation (2.11), the *basis set*. This set of functions must be chosen based on two conflicting requirements. On the one hand, it must be as large as possible, i.e. contain as many different functions as possible, so that the molecular orbitals constructed from combinations of the different basis functions ϕ can be as close as possible to the 'true' molecular orbitals. On the other hand, it must be as small as possible, so that the Hartree–Fock calculation will remain computationally tractable. Both of these requirements apply also to calculations using the methods to be described in Chapter 3, since molecular orbitals are used in these methods also.

A priori, many possible types of function could be used as the basis functions ϕ. However, it is found that *atomic orbital*-like functions, centred on the nuclei of the atoms making up the system, represent an *efficient* choice: it leads to quite accurate results with relatively small numbers of functions. In some calculations, other choices are made, but we will focus here on the atom-centred orbitals basis sets.

For hydrogenic atoms (the one-electron atom systems H, He^+, Li^{2+}, and so on), the Schrödinger equation can be solved analytically so the atomic orbital functions are known exactly. They are given by the following expression:

$$\phi_{nlm}(r,\theta,\varphi) = NY_{lm}(\theta,\varphi)P(r)e^{-\zeta r} \tag{2.15}$$

In equation (2.15), n, l, and m are the principal, angular, and magnetic quantum numbers, and r, θ, and φ are the spherical polar coordinates of a point with respect to the nucleus. N is a normalization constant, Y_{lm} is a spherical harmonic function, $P(r)$ is a polynomial function, and ζ (Greek letter *zeta*) is a number defining how steeply the function declines away from the nucleus, usually simply called the *exponent* in this context. For the simplest atomic orbitals, there are simple expressions for the spherical harmonic functions in terms of the Cartesian displacements x, y, and z with respect to the nucleus: for $l = 0$, $Y = 1$ (an s function); for $l = 1$, $Y = x, y$, or z (p_x, p_y, or p_z, functions); for $l = 2$, $Y = xy, xz, yz, x^2-y^2$ or $2z^2-x^2-y^2$ ($d_{xy}, d_{xz}, d_{yz}, d_{x^2-y^2}$ or d_{z^2} functions). The radial polynomial functions are also simple, of the form $P(r) = 1$ (1s and 2p functions), $P(r) = 1-kr$ (2s and 3p functions, where k is a constant), and so on.

Basis functions of the form in equation (2.15) (sometimes called *Slater* functions) are not very commonly used in quantum chemical calculations, as evaluating the integrals of (2.12) and (2.13), needed to construct the Fock operator, is very computationally demanding. Instead, *Gaussian* functions as in equation (2.16) are used. The corresponding integrals have an analytical closed form and can be evaluated much more readily.

$$\phi_{nlm}(r,\theta,\varphi) = NY_{lm}(\theta,\varphi)P(r)e^{-\zeta r^2} \tag{2.16}$$

The switch from the functional form of equation (2.15) to that of equation (2.16) looks small, but the square in the exponent means that Gaussian functions have a very different shape than the exponential Slater functions. This is shown in Figure 2.6 for a simple 1s function. The regular Slater function is plotted, together with a Gaussian function for which the exponent factor ζ and the normalization constant have been adjusted to yield the best overlap with the Slater function. Even so, it can be seen that the Gaussian function has the wrong shape for $r = 0$ (the exponential function has a sharp peak or cusp, whereas the Gaussian is smoothed off) and for large r (the Gaussian drops too quickly to zero).

At least for functions of inner atomic orbitals, which have significant density at or near the nucleus, Gaussian functions are not usually used in the simple form shown above. Instead, each basis function is itself a *combination* of several elementary (or *primitive*) Gaussian functions G_i each with a different exponent ζ_i, as in equation (2.17). The resulting combination is referred to as a contracted Gaussian basis function.

$$\phi = \sum_i^{n_{contr}} d_i G_i(\zeta_i) \tag{2.17}$$

In this equation n_{contr} is the number of primitive Gaussian functions used in the contracted function. The coefficients d_i give the weight of each primitive Gaussian in the

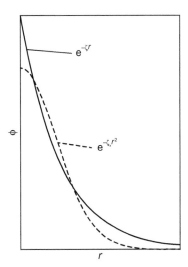

Figure 2.6 Shape of Slater and Gaussian functions.

contracted basis. The coefficients and exponents can be chosen based on calculations on simple systems, such as atoms, with a large basis set composed only of primitive Gaussian functions. In this way, the contracted basis function will have a shape more like that of a Slater function. Even with large n_{contr}, the integrals that need to be evaluated in order to construct the Fock operator are much more straightforward to calculate than with Slater basis functions. Figure 2.7 shows that even with only two primitive Gaussian functions, it is already possible to obtain a closer fit to a Slater function, though closer inspection shows that there is still a significant mismatch near the nucleus.

As written above, a good choice of basis set must combine computational efficiency and accuracy. A first logical choice that one might wish to make is to choose

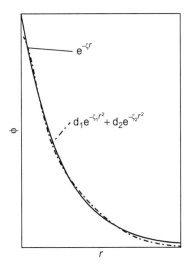

Figure 2.7 A contracted basis function, based on two primitive Gaussian functions, compared to a Slater function.

Minimal basis sets

A minimal basis set was defined earlier as the smallest basis set that could *possibly* be used to obtain reasonable molecular orbitals for the molecule. In some cases, there are occupied atomic orbitals in some of the neutral atoms making up a molecule which are not to any significant extent occupied in the molecule. Consider a cluster Na_nCl_n, i.e. a nanocrystal of sodium chloride. Due to the highly ionic nature of bonding in this system, none of the occupied molecular orbitals of the nanocrystal will have significant contributions from the sodium $3s$ orbital, and this orbital could be omitted from the basis set without catastrophic consequences, even though it is of course occupied in the neutral sodium atom.

a *minimal basis set*. This is a basis set containing one (contracted Gaussian) basis function for every occupied atomic orbital in the atoms making up the system being studied. This would appear to be the bare minimum that is needed to construct the molecular orbitals. In fact, such basis sets can be chosen and do give qualitatively reasonable results. The basis functions can be chosen e.g. to reproduce the accurate shape of the atomic orbitals in Hartree–Fock calculations on the atoms. A common choice is to assign a fixed number of primitive Gaussian basis functions to each of the atomic orbitals. Like for other basis sets, compilations of contraction coefficients and exponents have been assembled for each of the different atoms in the periodic table, and can be consulted in repositories. Minimal basis sets are also inbuilt into most software used to perform Hartree–Fock calculations, where their use can be specified with a simple keyword. A common basis set of this type is the STO-3G basis set. The acronym means 'Slater Type Orbital, based on contractions of 3 primitive Gaussian functions'. This basis set uses Gaussians chosen to emulate a certain set of Slater orbitals as closely as possible.

In quantitative terms, minimal basis sets tend to yield very poor results. This can be understood based on the combined effects of *orbital breathing* and of *orbital polarization*. Orbital breathing refers to the fact that atomic orbitals do not have fixed sizes: they can expand or contract in response to the bonding environment, and in particular to the flow of electrons in the system. Accumulation of electrons in an orbital will cause it to expand, and vice versa. For example, the $1s$ orbital in a hydrogen atom is more compact than the $1s$ orbital in a hydride anion H^-. Likewise, the hydrogen $1s$ orbital contributions to the appropriate molecular orbitals will be progressively more compact on going from molecular LiH to CH_4 to H_2O. In order to represent this effect, the basis set needs to contain at least *two* basis functions for each occupied atomic orbital in the atoms making up the system. The 'tighter' of the two basis functions will contribute more to molecular orbitals when the underlying atomic orbital adopts a more compact form, and the 'looser' or more diffuse of the two basis functions will contribute more when the underlying atomic orbital is more expanded. A basis set of this type is referred to as being a *double-zeta* (DZ) basis set. The name derives from the Greek letter ζ, often used as the symbol for the exponent factor in equations (2.15)–(2.17), which determines the 'tightness' or 'looseness' of a basis function. By contrast to double-zeta basis sets, minimal basis functions are sometimes referred to as single-zeta basis sets.

In practice, inner or core atomic orbitals do not 'breathe' much in this way, so a fully double-zeta basis set is not usually needed, with only the valence atomic orbitals requiring representation by two basis functions. This leads to so-called valence double-zeta (VDZ) basis sets, also called *split-valence* (SV) basis sets. Still greater accuracy can be obtained by using triple-zeta (or valence triple-zeta) basis sets. Provided the basis functions are well chosen, going much beyond triple-zeta for Hartree–Fock calculations is usually not necessary in the sense that it does not lead to improved predictions of experimental properties. One exception concerns the provision of appropriate basis functions to describe the very diffuse electron density present in species with negative charge (or significant local negative charge, in the case of polar compounds). As basis sets are often developed through calculations on neutral atoms, this diffuse density is not always well described by standard basis sets, even if they are of triple-zeta quality. Augmenting the basis set with additional diffuse functions is then needed.

Atomic orbitals can also polarize. The basic atomic orbitals of equation (2.16) have a definite shape, dictated by the spherical harmonic part. For example a *p* function has two lobes, which are of necessity aligned exactly with an axis passing through the nucleus. An appropriate combination of the p_x, p_y, and p_z orbitals can produce a *p* orbital with an arbitrary direction, but it will always be oriented along a given linear axis. In a non-symmetric molecular environment, better bonding (e.g. π bonding) may result if the lobes can instead 'tilt' one way or another, thereby 'polarizing' the orbital. Even a multiple-zeta basis set does not permit this. In order to enable polarization, one needs to include additional basis functions, of different angular quantum number compared to the orbital one wishes to allow to relax. For example, as shown here (see margin), slight admixture of *d* orbitals can allow *p* orbitals to 'tilt' or polarize. Likewise, *p* functions can polarize *s* functions. Polarization occurs more readily for 'softer' (more polarizable!) atoms, so it is a more important effect for third-row elements such as S or P compared to second-row O or N. Still, this effect is important enough that polarized basis sets need to be used to obtain accurate results even for first- and second-row elements.

As mentioned above, the typical user of a quantum chemical software package does not need to develop their own basis sets. This is a specialized task, which requires attention to computational efficiency as well as all the factors mentioned above. Instead, experts have developed various 'standard' sets of basis functions for individual atoms, which are made available through repositories or through the use of keywords in the software package. A basis set of the same size and developed in roughly the same way will often be developed for many atoms at the same time. Such families of basis sets are then given a name or acronym which allows the expert user to assess the properties of the basis set without having to consider in detail its parameterization. There are many of these acronyms, and their diversity can be a bit bewildering for a beginner!

It is usually sensible to construct the overall basis set for a molecular system by using basis functions centred on each atom that belong to the same family of basis sets. There is little benefit to treating one atom with a triple-zeta basis set if the neighbouring atom is treated using a minimal basis. Instead, one should try to use a *balanced* basis set. However, in large molecules, it can be useful to describe some atoms with larger numbers of basis functions than others. For example, in a transition metal complex, one might choose a bigger basis set for the metal and the atoms directly coordinated to it, and a smaller one for the (less important) atoms situated around the periphery.

While using an insufficiently large basis set obviously leads to an incorrect description of the molecular orbitals, it also leads to errors in calculated energies. One way in which this can occur is when computing interaction energies between two fragments A and B through calculation of the energy of A, of B, and of the interacting system AB. All three calculations will yield energies that are less negative than would be obtained with infinitely large basis sets. However, the error may be a bit smaller for AB, as the basis functions located on atoms of A may overlap with the region where the electronic wavefunction of B is located, and can contribute to the description of the orbitals of B in the AB system. This effectively increases the size of the basis set on B in the AB calculation, leading to a smaller error than in the isolated B, and a spurious contribution to the interaction energy. This effect known as **basis set superposition error** can be corrected by carrying out the fragment calculations using the extended basis set of the whole system (the so-called counterpoise correction).

For calculations on systems containing atoms with a large atomic number, there are many inner core electrons and many core orbitals. As mentioned above, these orbitals do not change shape very much from one species to another, yet their presence still adds to the computational challenge of describing the system. Also, the

---

Some common basis set families

The STO-3G minimal basis sets have been mentioned in the main text. Several other series of basis sets were put forward by the group of Nobel Prize winner John Pople in the 1970s and 1980s and remain in common use. The next smallest member of this family is referred to as 3-21G; it is a split-valence basis set. For atoms of the second row, such as carbon or oxygen, it contains one s basis function corresponding to the $1s$ core, which is a contracted basis function as in equation (2.17) constructed from three primitive Gaussians ($n_{contr} = 3$). For the valence orbitals, there are 2 s functions, and 2 p_x, p_y, and p_z functions. The tighter of each of these is a contracted basis function with two Gaussian primitives, whereas the more diffuse one is a single primitive Gaussian function. Hence the 3-2-1 labels in the acronym. The 6-31G basis is constructed on the same split-valence pattern, but with six primitives for the core orbitals and three and one, respectively, for the valence orbitals. The 6-311G basis set is of triple-zeta size. These basis sets are accompanied by diffuse and polarization functions, with a range of notations used to signal their inclusion. For example, the 6-311+G(d,p) basis set includes additional diffuse functions on all atoms except hydrogen (it is therefore in some sense a quadruple-zeta basis set), as well as d polarization functions on these 'heavy' atoms, and p polarization functions on hydrogen atoms.

Another commonly used family is the split-valence, triple-zeta, and quadruple-zeta family denoted 'SVP', 'TZVP' and 'QZVP'. A second revision of these basis sets was issued that improved their performance, and these are known as def2-SVP, def2-TZVP, and def2-QZVP.

A third widely used family of basis sets is that of the correlation-consistent basis sets, in their original form written as cc-pVXZ, where X is D for the double-zeta basis set, T for triple-zeta, Q for quadruple-zeta, and so on. These basis sets include extensive sets of polarization functions and where needed a form augmented with diffuse functions (aug-cc-pVXZ) is available.

electrons in these orbitals have a mean velocity that is a significant fraction of the speed of light, so that they begin to display relativistic effects. While effects of relativity are largest for the innermost core of $1s$ electrons, the other s and indeed other p, d, and f orbitals do experience knock-on effects that change the overall properties of the system. To understand such chemical effects of relativity requires in principle a treatment going beyond the normal non-relativistic Schrödinger equation. There are quantum chemical methods that explicitly include relativistic effects, though these will not be discussed here.

For heavy atoms, though, there is a way to choose a 'basis set' in such a way as to reduce the computational complexity and, where needed, to describe at least roughly the relativistic effects: this is to use an *effective core potential* or ECP. The idea is to remove the innermost core electrons, e.g. the $1s$, $2s$, and $2p$ electrons for a fourth-row transition metal centre such as Fe, transforming it from a centre with 26 electrons to one with only 16. The charge on the core is also modified. Simply removing the innermost 10 electrons and adjusting the charge of the nucleus would merely change an iron atom into a sulphur atom. Instead, the functional form of the interaction between the remaining electrons and the Fe^{16+} core is modified also, so as to mimic the effect that the 10 core electrons would have, and at least roughly reproduce the wavefunctions that would be obtained when carrying out calculations with all electrons. Many families of ECPs are available, parameterized to reproduce the results of all-electron calculations. Some ECPs are fitted to relativistic calculations.

Some of the sophisticated methods that have been developed to make Hartree–Fock and other quantum chemical calculations faster to run require so-called *auxiliary basis sets*. These are sets of fixed functions, chosen prior to running the calculations, and which are used as 'building blocks' to assemble more complex functions generated during the calculation. Just as the usual atomic orbital basis set of equation (2.11) is used to construct the molecular orbitals, these additional auxiliary basis functions are used to assemble other functions needed to carry out the calculation efficiently. One example is the products of basis functions present in equation (2.14)—the product of two functions is similar to a density, and choosing a set of auxiliary basis functions to represent these products or densities as closely as possible is accordingly sometimes referred to as *density fitting*. The precise roles of auxiliary basis sets go beyond the scope of this book. As for the main basis set, though, the user of a quantum chemical computer program may need to choose these auxiliary basis sets in order to run calculations. Broadly speaking, larger sets of auxiliary basis functions will lead to more demanding but more accurate calculations.

2.7 Further reading

- *Molecular Quantum Mechanics*, Peter W. Atkins and Ronald S. Friedman, 5th edition, Oxford University Press, Oxford, 2010. This book gives a good general introduction to quantum mechanics and its implications for chemistry.

- *Modern Quantum Chemistry: Introduction to Advanced Electronic Structure Theory*, Attila Szabo and Neil S. Ostlund, Dover Publications, Mineola, New York, 1996. This book, while old (it is a reprint by a different publisher of the 1989 second edition; the first edition is from 1982), provides a detailed yet accessible introduction to quantum chemistry, in particular the Hartree–Fock method.

- *Quantum Chemistry*, 7th edition, Ira N. Levine, Pearson, 2013. This book provides a general introduction to quantum mechanics and its applications to chemistry, with detail of mathematical derivations.

2.8 Exercises

2.1   Using a quantum chemical code, carry out your own Hartree–Fock calculation on a simple molecule, such as water. Identify the total energy of equation (2.7) and the molecular orbital coefficients of equation (2.11). Also identify the sections of the output from the program that describe the way in which the initial guess for the orbitals was generated, and which describe the procedures used to accelerate convergence.

2.2   Using a quantum chemical code and a graphical viewer for molecules, calculate a Hartree–Fock wavefunction then visualize isovalue surfaces for the molecular orbitals of a simple molecule such as formaldehyde, $H_2C=O$. Also identify the molecular orbital coefficients of equation (2.11) and check that these correspond to the graphical representation.

2.3   Carry out Hartree–Fock calculations using the 6-311G basis set for a hydrogen atom and a hydride anion; compare the contributions made to the occupied orbital by the three basis functions and hence identify the effect of orbital breathing.

2.4   For the ethene molecule C_2H_4, carry out Hartree–Fock calculations using the 6-31G and 6-31G(d) basis sets. Inspect the output to check the details of the basis set used. Compare the overall energies (equation (2.7)) obtained, and also identify the magnitude of the contribution from the *d* polarization basis function on the carbon atoms to the different atomic orbitals.

2.9 Summary

- The Schrödinger equation describes all aspects of molecular behaviour.
- Solution of the time-independent, electronic, Schrödinger equation, gives the potential energy of a molecular system for a given arrangement of the atoms.
- Solving the electronic Schrödinger equation analytically is not possible, and for most systems, exact solutions are not accessible even using numerical methods.
- Instead, various approximate techniques are used.
- The central computational method for solving the electronic Schrödinger equation is the Hartree–Fock method, in which an approximate wavefunction is obtained, written as an antisymmetrized product of one-electron functions, the molecular orbitals.
- Carrying out a Hartree–Fock calculation requires some understanding of the role of the *initial guess* for the orbitals, of techniques for accelerating calculations, and of the nature of atomic orbital basis sets.
- While the Hartree–Fock method is nowadays seldom used to produce research results, it is frequently used as the first step in other quantum chemical studies, and many other methods bear strong similarities to it.

3 Quantum Chemical Methods

3.1 Introduction

The remarkable success of quantum chemical methods in contemporary chemical research is due to the availability of powerful computers, of course, but especially to the development of accurate methods and of efficient algorithms that implement these methods. The previous chapter introduced the quantum mechanical framework within which electronic structure theory is developed. It then gave a rather detailed description of the approximations involved in Hartree–Fock theory, the detail of how Hartree–Fock calculations are performed, and the nature of the results obtained. In this chapter, we will consider more accurate and hence more practically useful quantum chemical methods. Some of these methods build upon the results of a Hartree–Fock calculation as a first step, while others bear strong similarities to Hartree–Fock, so having a sound understanding of what is involved in a Hartree–Fock calculation is essential for performing quantum chemical calculations. Hence this chapter is not supposed to be read in isolation, but rather should be studied together with Chapter 2.

3.2 Correlated *ab initio* methods

Hartree–Fock theory is an approximate theory, because it is based upon the calculation of an approximate wavefunction, the Slater determinant of equation (2.5). Because of the variational principle, the Hartree–Fock energy is therefore higher (less negative) than the true ground state energy. The difference between the two, $E_{exact} - E_{HF}$ (a negative quantity), is called the *correlation energy*. The physical reason for this difference is that electron positions are *correlated*: the probability that a given electron will be in one place is dependent on where another electron is placed at the same moment. For systems containing only one electron, such as the H atom, He^+ ion, or H_2^+ ion, the Hartree–Fock method is exact (at least within the Born–Oppenheimer approximation, and within the limitations of the basis set used). But for all other systems, methods that take into account the correlation effect are needed in order to obtain accurate results.

Same-spin and opposite-spin correlation effects

What is written above is not exactly correct: the Hartree–Fock wavefunction *does* account for some aspects of correlation, namely the correlation in the positions of electrons of *same spin*. Antisymmetrization of the wavefunction, as in the Slater

(continued...)

determinant of equation (2.5), does not introduce a major change compared to the simple product formulation of equation (2.3). But there is a difference between the two expressions! By requiring that the wavefunction change sign when the coordinates of two electrons are switched (equation (2.4)), then by necessity the wavefunction must be equal to zero for configurations where two electrons have the same coordinates. Two electrons of opposite spin differ in their spin coordinate, so this effect has no impact in that case. But for a configuration where two electrons of *same* spin are in the same position, the Hartree–Fock wavefunction is exactly zero. For regions where two electrons of same spin are close to one another, continuity of the wavefunction also means that the wavefunction has a small magnitude. In this sense, the Hartree–Fock wavefunction is 'correlated' in some sense (it is sometimes said to describe *Fermi correlation*). However, whenever the word 'correlation' is used in quantum chemistry, what is meant is all the effects that go beyond Hartree-Fock theory, as described by correlated methods.

The two standard methods to describe correlation can be introduced by restating the key approximation in Hartree-Fock theory in two different ways. These are based on considering either equation (2.7) or equation (2.10) as a starting point. In the logic of equation (2.7), the Hartree–Fock method calculates the correct energy expectation value, calculated with the correct $\hat{H}_{elec}$ Hamiltonian operator, but with an approximate wavefunction (of the form of equation (2.5)). The variational principle ensures that the wavefunction is as similar as possible to the exact solution, within the limitations of the molecular orbital approximation. In the logic of equation (2.10), on the other hand, the Hartree–Fock method returns the exact wavefunction but as a solution to an approximate one-electron problem, the Fock equations. In the next sections, we will consider both approaches and the associated methods that they lead to.

3.3 The variational approach: configuration interaction

The configuration interaction method can be introduced with reference to equation (2.7). By using a more flexible approximate form for the wavefunction than in Hartree-Fock theory, and combining this more accurate form with the variational method, better results can be obtained. In the Hartree-Fock method, a Slater determinant wavefunction (equation 2.5) is used, which has the character of an (antisymmetrized) *product* of one-electron functions or molecular orbitals. The true wavefunction has an intrinsically many-electron character that cannot be described in the product formalism. In principle, there are many ways to build up a trial wavefunction with many-electron character. However, it is also important to do so in a way that is computationally efficient. In practice, the most convenient procedure to build many-electron functions is to take a linear combination of many Slater determinants, including the Hartree-Fock determinant. While one Slater determinant alone does not describe correlation, such linear combinations do.

Solving the Fock equations generates as many spatial molecular orbitals as there are atomic orbital basis functions in the basis set. For a closed-shell system with n electrons described using a restricted Hartree-Fock wavefunction, the $n/2$ lowest-energy of these orbitals are *occupied*, whereas the remaining spatial orbitals are empty

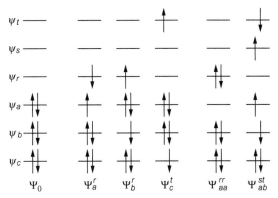

Figure 3.1 Schematic representation of Slater determinants involved in configuration interaction wavefunctions.

or *vacant*. As shown in Figure 3.1, it is possible to generate from the occupied and vacant orbitals a large number of other Slater determinants, by moving electrons from the occupied orbitals to vacant ones. The occupied orbitals are labelled ψ_a, ψ_b, ψ_c and so on by decreasing order of energy, while the vacant orbitals are labelled ψ_r, ψ_s, ψ_t, ... by increasing energy. With reference to the Hartree–Fock Slater determinant Ψ_0, there are different types of additional determinants. The determinant Ψ_a^r shown in Figure 3.1 is an example where *one* electron has been moved from occupied orbital ψ_a to vacant orbital ψ_r. This is a *singly excited* determinant (or a single excitation). Two other singly excited determinants, and two doubly excited determinants (or examples of double excitation) are also shown. For a system with n electrons, it is possible to have up to n-tuple excitations. The most general expression for the resulting wavefunction will be a variationally determined combination of all possible determinants of all possible levels of excitation from 1 to n, and can be written as follows:

$$\Psi_{CI} = C_0\Psi_0 + \sum_{a,r} C_a^r\Psi_a^r + \sum_{a,b,r,s} C_{ab}^{rs}\Psi_{ab}^{rs} + \sum_{a,b,c,r,s,t} C_{abc}^{rst}\Psi_{abc}^{rst} + \dots \tag{3.1}$$

In this expression, the summations are over all possible excitations of the given type, with the first summation corresponding to all possible single excitations (a and r refer here to any possible occupied and vacant orbital, not a specific one as in Figure 3.1), the second summation being over all double excitations, and so on. The coefficients C_0, C_a^r, and so on, measure the weight of the corresponding determinant in the wavefunction, and it is these coefficients that are optimized using the variational principle. If all possible terms in equation (3.1) are considered, one has the so-called Full Configuration Interaction (FCI or full CI) method, and this yields the *exact* energy for the electronic ground state (and indeed the excited states), within the limitations of the basis set used. By including multiple determinants, one can thereby produce the exact solution to the Schrödinger equation!

The small admixture of excited determinants changes the energy and it also changes the *wavefunction* quite significantly. As discussed in Chapter 2, there are many ways to plot the wavefunction, which in the case of the dihydrogen molecule is a function of six variables. In order to visualize the difference between the Hartree–Fock and CI wavefunctions, it is helpful to make a plot of Ψ^2 for a small part of this six-dimensional space, where the difference is relatively large: we consider only the coordinates where

> ## FCI versus HF energy
>
> The FCI energy for molecules is much more accurate than the HF energy. Consider H_2, described using the STO-6G minimal basis, at a H–H distance of 0.75 Å. The HF energy is -1.12473 hartree, whereas FCI gives -1.14574 hartree, 0.02101 hartree or more than 50 kJ mol^{-1} lower. In this small basis set, there is only one doubly excited determinant included (corresponding to double occupation of the σ^* antibonding orbital), and the variational coefficient obtained is -0.115, vs 0.993 for the Hartree–Fock (or reference) determinant. Given that the *square* of the coefficients measures the approximate weight of the different determinants, this means that a roughly 1% admixture of the excited determinant is enough to lower the energy by 50 kJ mol^{-1}.

both electrons lie along the molecular axis, and plot Ψ^2 for different values of z_1 and z_2, the distances of each electron along the internuclear axis. The value $z = 0$ corresponds to the centre of the bond, while the nuclei are positioned at $z = \pm 0.375$ Å. The HF wavefunction, using the cc-pV5Z basis, yields the Ψ^2 shown in Figure 3.2. The wavefunction is large in amplitude when one electron is at the position of one of the nuclei, and the other is at the position of the *other* nucleus, but it is also large in amplitude when *both* electrons are at the position of one of the nuclei. Hence there are four peaks in Ψ^2. This is a typically *non*-correlated wavefunction.

In contrast, consider a CI wavefunction in which just one doubly excited configuration, that in which both electrons are excited to the σ^* orbital, is included. This yields the plot for Ψ^2 shown in Figure 3.3. There are still four peaks, but they are now of different size. The peaks for $z_1 = -0.375$, $z_2 = 0.375$ and vice versa (those corresponding to electrons near distinct nuclei) are now much bigger than those for $z_1 = z_2 = -0.375$ and $z_1 = z_2 = 0.375$ (electrons both near the same nucleus). In fact, the peaks with $z_1 = -z_2$ are now higher than they were in Figure 3.2, whereas those with $z_1 = z_2$ are lower. In general, the whole region lying along the diagonal line $z_1 = z_2$ has a much smaller amplitude than in the uncorrelated wavefunction of Figure 3.2.

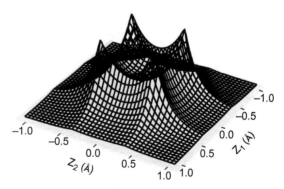

Figure 3.2 Three-dimensional plot of $\Psi^2(z_1, z_2)$ for the HF/cc-pV5Z wavefunction of the hydrogen molecule.

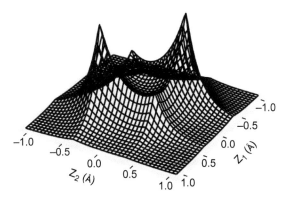

Figure 3.3 Three-dimensional plot of $\Psi^2(z_1,z_2)$ for the minimal CID/cc-pV5Z wavefunction of the hydrogen molecule. The vertical scale used is the same as in Figure 3.2.

Unfortunately, full CI calculations are not possible except for relatively simple systems. There are simply too many ways to spread the electrons in the orbitals—the number of Slater determinants in equation (3.1) increases exponentially with the number of electrons in the system and the number of basis functions. However, it is possible to carry out CI calculations using a truncated version of equation (3.1), in which only some of the possible excitations are included. Such methods scale less steeply with the system size, though it remains the case that doubling the size of the molecule to be treated will increase the computer time and storage needed by at least a factor of 32 (i.e. 2^5) when using standard computer codes. Accurate results including correlation effects for large systems remain challenging, though much work has been done in recent years to develop modified routines that scale less steeply with system size.

One type of truncation of equation (3.1) is relatively trivial to make and does not lead to a substantially less accurate result. For example, the electrons that occupy the inner cores of the atoms (e.g. $1s$ orbitals for second-row atoms such as C or F, $1s$, $2s$, and $2p$ for third-row atoms such as Si or S) can be omitted from the CI expansion. This is not because their motions are not correlated with each other, or with those of the valence electrons. Rather, it is because the portion of the correlation energy due to such effects will be pretty much the same in any molecule containing those atoms, so that when considering *relative* energies, they will almost exactly cancel out. This 'frozen-core' approximation is very commonly used in correlated calculations.

It is also possible to truncate the full CI expansion of the wavefunction by considering only a subset of the possible classes of excitations. The most important excitations, those that lead to the greatest variational lowering of the energy, are the *double* excitations (in hand-waving terms, this is because correlation is a physical phenomenon involving two electrons at a time). This means that the simplest CI method is so-called 'configuration interaction with double excitations', CID. Single excitations do not in themselves help in describing correlation (so the CIS method returns the same energy as HF), but including them slightly modifies the coefficients obtained for the double excitations, and there are many fewer single excitations than double ones, so a more commonly used method is CI with single and double excitations, CISD. This is a much more demanding calculation than HF, but much less than FCI.

Double excitations

The dominant role of double excitations can be seen by considering the total energy obtained in a set of truncated CI calculations on the water molecule (using the cc-pVDZ basis set, an O–H distance of 1.0 Å, and an H–O–H angle of 100°) as shown in the table. The HF energy is −76.02129 hartree. For this small system with only 19 vacant orbitals and only 8 electrons needing correlation (the electrons in the O $1s$ orbital are not included in the CI expansion), it is possible to carry out FCI calculations, as well as calculations with truncation to single and double excitations (CISD), single, double, and triple excitations (CISDT) and single, double, triple, and quadruple excitations (CISDTQ). Over 90% of the overall FCI correlation energy is recovered by CISD, with the bulk of the remaining energy being obtained with CISDTQ (CISDT does not have much effect relative to CISD).

Level	E (/hartree)	E_{corr} (/hartree)
HF	−76.02129	0.00000
CISD	−76.22749	−0.20620
CISDT	−76.23066	−0.20937
CISDTQ	−76.23970	−0.21841
FCI	−76.24006	−0.21877

The CI method is very useful for understanding electron correlation, but it is in practice seldom used, because it has one severe flaw: it is *size-inconsistent*. This property, and its opposite, *size-consistency*, can be readily understood by considering the hydrogen molecule example given above. Imagine that instead of one H_2 molecule described by HF and CI, we had two, 100 Å apart but with the same internal structure. At the HF level (and the same is true with many other approximate quantum chemical methods), the overall energy is simply twice the energy of the single H_2, −2.24946 hartree. With CISD, we would expect to obtain an energy of twice −1.14574, i.e. −2.29148 hartree. In fact, the calculation yields an energy of −2.29094 hartree, higher by 0.00054 hartree. This is not a huge error (about 1.5 kJ mol^{-1}), but the error increases with system size (an analogous calculation for two distant water molecules as in the other example yields an error of over 50 kJ mol^{-1}), so it is an important effect. It makes it impossible to obtain sensible relative energies by comparing the energy of two separate fragments and the species formed by combining them—something that one often wishes to do in computational studies. The error arises due to the truncation to single and double excitations: it means that the wavefunction obtained for the larger system is not exactly the same as the product of the wavefunctions for the separate molecules. It is easy to see that this product would comprise quadruple excitations. Although there are (approximate) ways to account for this error (e.g. the Davidson correction, introduced by theoretical chemist Ernest Davidson), its existence means that CISD is not, in practice, much used.

3.4 The perturbation approach: Møller–Plesset theory

To understand the perturbation approach, it is useful to consider the set of equations (2.10), the Fock equations, as being a set of equations that resemble the original Schrödinger equation, but with an approximate Hamiltonian operator, the Fock operator f. The advantage to thinking this way is that unlike the normal Schrödinger equation, the Fock equations can be solved *exactly* (in a self-consistent way, and using a finite basis set), yielding the orbitals ψ_i and their energies ε_i as well as the overall Hartree–Fock wavefunction Ψ_0. In quantum mechanics, if one knows the *exact* solution to the Schrödinger equation for a given Hamiltonian operator, one can write out a series of successively closer approximations to the solution for a *different* Hamiltonian, provided that this other Hamiltonian only differs from the first one by some small perturbative amount. This approach to solving the Schrödinger equation is known as perturbation theory and can be applied in many contexts.

For quantum chemistry, this is applied by writing the exact Hamiltonian as a sum of the approximate Hamiltonian (here $\hat{f}$, or more precisely the closely related $\hat{F}$, the many-electron Fock operator, which is simply a sum of one-electron Fock operators $\hat{f}$) and a small 'perturbation' ($\hat{V}$) which is the difference between $\hat{H}$ and $\hat{F}$, multiplied by a constant, λ (which is convenient to include for reasons that will appear below, though its value is simply 1).

$$\hat{H} = \hat{F} + \lambda(\hat{H} - \hat{F}) = \hat{F} + \lambda\hat{V} \tag{3.2}$$

By inserting the expression in equation (3.2) into the Schrödinger equation, and carrying out some manipulations, one obtains expressions for the wavefunction and for the energy that are approximate solutions to the Schrödinger equation with the correct Hamiltonian, expressed in terms of the exact solutions to the Fock equations. The expression for the energy is written as a sum of terms depending on successive powers of λ:

$$E_{\text{exact}} = E_0 + \lambda E_1 + \lambda^2 E_2 + \lambda^3 E_3 + \cdots \tag{3.3}$$

In equation (3.3), the (complicated) expressions for E_1, E_2, E_3 have not been written out in detail, but they can be obtained from the solutions to the Fock equations, the expression for the Fock operator, and the Hamiltonian operator. Using λ in equation (3.2) has the advantage of grouping together in equation (3.3) the terms that depend strongly on the small correction term in (3.2), which collectively equal E_1, those which depend a bit less strongly on it, which equal E_2, and so on. E_0 is the energy corresponding to the solution for the approximate operator $\hat{F}$, E_1 is the first-order perturbation theory correction, E_2 the second-order correction, and so on. Perturbation theory was first applied to chemical systems by the physicists Møller and Plesset, so it is often called Møller–Plesset theory in this context, and the energy $E_0 + E_1 + E_2$ is referred to as the MP2 energy. MP3, MP4, and so on refer to energies including the next correction terms also.

Because the perturbation is supposed to be small (remember, the Hartree–Fock solution is quite close to the exact wavefunction, as discussed above), these corrections should become smaller and smaller, so the MPn energy should converge on the exact energy as n becomes larger. Because of the way that the perturbation operator was chosen in equation (3.2), the Hartree–Fock energy turns out not be

E_0, but $E_0 + E_1$, so the first substantive energy correction term is E_2. This term can be calculated based on the Hartree–Fock solution, and involves a sum of many contributions.

The energy correction in MP2 bears some resemblance to the terms that lead to an energy lowering by double excitations in the CISD method. This is not a coincidence, because there are underlying parallels in the ways that CISD and MP2 describe correlation. Hartree–Fock can be *described* either as a method based on optimizing an approximate wavefunction with an exact Hamiltonian, or as a method based on obtaining the exact solution to an approximate Hamiltonian. But in reality it is both of these. Configuration interaction and perturbation theory both set out to yield an improved wavefunction and an improved energy, so it is not surprising that somewhat similar terms arise in both.

There are, however, significant differences. Configuration interaction is a variational method, so it always yields an energy higher than the exact ground-state energy. The Møller–Plesset perturbation theory approach is not variational, and it can yield energies lower than the true energy in some cases. This occurs especially when the energy gap between the Hartree-Fock occupied and vacant orbitals is small, so care is needed when using MP*n* methods in this case. It should also be noted that the MP*n* energy series does not always converge smoothly towards the FCI energy with increasing *n*, so that the use of MP*n* methods with large *n*, which are quite computationally demanding, is not always particularly recommended. Another difference is that MP*n* methods are size-extensive, unlike CISD and other truncated CI methods.

A closer look at MP*n* methods

As above, it is instructive to look at an example. For the same water molecule as above, with the same basis set, the MP2 energy is −76.22653 hartree, fairly close to the CISD energy. With MP3 and MP4, it is respectively −76.23305 and −76.23893 hartree, with the latter value rather close to the FCI value of −76.24006. The perturbation approach yields a reasonable estimate of correlation effects for systems with well-separated HOMO and LUMO orbitals, such as the water molecule near its equilibrium structure.

3.5 Coupled-cluster methods

Coupled-cluster theory is an alternative approach to describe correlation. As in configuration interaction theory, this approach relies on a generalized wavefunction, written as a combination of Slater determinants. The expansion, written as equation (3.4), differs in appearance quite strongly from that of CI (equation (3.1)), as it involves an *excitation operator T* and an exponential term:

$$\Psi_{CC} = Ne^{\hat{T}}\Psi_0 = N\left(1+\hat{T}+\frac{\hat{T}^2}{2}+\frac{\hat{T}^3}{6}+\dots\right)\Psi_0. \tag{3.4}$$

In this equation, N is a normalization constant. The excitation (or cluster) operator $\hat{T}$ acts on the Slater determinant to introduce excited configurations (the same ones as in CI theory) with expansion coefficients. Due to the exponential notation (shown above also as its Taylor expansion), the coupled-cluster wavefunction Ψ_{CC} includes contributions from the action of $\hat{T}$ on the Hartree–Fock wavefunction, but also contributions from the action of the successive powers of this operator, $\hat{T}^2$, $\hat{T}^3$, and so on. As in CI theory, it is necessary to truncate the order of the expansion in order to have a realistic method. This is done by writing $\hat{T}$ as a sum of terms $\hat{T}_1$, $\hat{T}_2$, $\hat{T}_3$, ... which introduce single, double, triple, ... excitations, and then keeping only the first few terms. The effect of the $\hat{T}_1$ operator is illustrated in equation (3.5):

$$\hat{T}_1 \Psi_0 = \sum_{a,r} t_a^r \left(\hat{\tau}_a^r \Psi_0 \right) = \sum_{a,r} t_a^r \Psi_a^r \tag{3.5}$$

$\hat{T}_1$ contains the effects of all possible single-excitation operators $\hat{\tau}_a^r$, which each correspond to exciting an electron from an orbital a, occupied in the Hartree–Fock reference, to a vacant orbital r. $\hat{T}_1$ also contains coefficients t_a^r (which are called *amplitudes* in coupled-cluster theory). In a similar manner, $\hat{T}_2$ generates all double excitations. Where only $\hat{T}_1$ and $\hat{T}_2$ are kept, one has the CCSD method, with the wavefunction given by equation (3.6):

$$\Psi_{CCSD} = N e^{\hat{T}_1 + \hat{T}_2} \Psi_0 \tag{3.6}$$

Again this expression can be rewritten using a Taylor expansion, which yields an infinite set of terms, a few of which are shown in equation (3.7):

$$\Psi_{CCSD} = N \left(1 + \hat{T}_1 + \hat{T}_2 + \frac{\hat{T}_1^2}{2} + \hat{T}_1 \hat{T}_2 + \frac{\hat{T}_2^2}{2} + \frac{\hat{T}_1^3}{6} + \dots \right) \Psi_0 \tag{3.7}$$

From this equation, it can be seen that the wavefunction will contain single excitations (from single application of $\hat{T}_1$), double excitations (from single application of $\hat{T}_2$, or double application of $\hat{T}_1$), triple excitations ($\hat{T}_1^3$ and $\hat{T}_1 \hat{T}_2$), and so on. This means that unlike CISD, where there are 'missing' quadruple excitations which lead to non-size-consistency, the coupled-cluster method contains all orders of excitation. As a result, coupled-cluster theory is size-extensive.

However because the wavefunction contains all orders of excitation, optimizing the amplitudes variationally has essentially the same computational demand as FCI, and as such would barely be applicable. Fortunately, an approximate scheme can be used to obtain the excitation amplitudes, and the coupled-cluster energy, and this approach is much less demanding. For example, CCSD theory, in which the excitation operator is truncated to $\hat{T}_{11}$ and $\hat{T}_2$, requires roughly the same computational effort as is needed for CISD.

As with CI, it is possible to truncate the coupled-cluster operator at several different levels, yielding the methods CCSD, CCSDT, CCSDTQ, ... upon including every term up to $\hat{T}_2$, $\hat{T}_3$, and $\hat{T}_4$ respectively. It is also possible to compute the approximate effect of including $\hat{T}_3$ on the CCSD energy, using perturbation theory. This can be done in several ways, the most common of which leads to results denoted CCSD(T). Likewise, one can include $\hat{T}_4$ approximately on top of CCSDT, yielding CCSDT(Q).

Results of coupled-cluster methods: an example

Using the same example of water near its equilibrium structure, the table below shows CCSD, CCSD(T), CCSDT, and CCSDT(Q) results for the water molecule with the cc-pVDZ basis set. HF and FCI results are included for reference. It can be seen that CCSD is much closer to FCI than to HF, and all the other methods are extremely close to FCI.

Level	E (/hartree)
HF	−76.02129
CCSD	−76.23605
CCSD(T)	−76.23935
CCSDT	−76.23953
CCSDT(Q)	−76.24007
CCSDTQ	−76.24003
FCI	−76.24006

3.6 Basis sets, correlation effects, and explicit correlation

One important aspect has been left out of the above discussion on correlated methods: the basis set. In section 2.6, the importance of the basis set for describing the molecular orbitals (or one-electron wavefunctions) was described. However, basis sets play a second role in calculations using correlated methods such as CISD, MP2, or CCSD(T): they provide the building blocks to construct the *many-electron* wavefunction. Because so much of the qualitative discussion of molecular electronic structure focuses on molecular orbitals, which are *one-electron* wavefunctions, the many-electron wavefunction is hard to visualize in an intuitive way for many people. Indeed, the graphical wavefunction representations most often used in chemistry are the orbital plots covered in section 2.4, which illustrate the molecular orbitals.

Yet it is the properties of the many-electron wavefunction, and in particular the way in which the wavefunction changes as two electrons move close to one another, that determine the correlation energy. Here it can be helpful to refer again to Figures 3.2 and 3.3, which illustrate part of the many-electron wavefunction for a very simple system containing just two electrons. In Hartree–Fock theory (Figure 3.2), the electrons' positions are not correlated, so the wavefunction does not change very strongly as one electron approaches another. The exact wavefunction is, however, much smaller in magnitude than the Hartree–Fock wavefunction when two electrons are close to one another (Figure 3.3). This decrease gets described through the population of excited determinants as in equation (3.1). The accuracy with which the decrease is described depends not only on the *method* used (e.g. CCSDTQ describes it more accurately than MP2), but also on the size of the basis set. Larger basis sets lead to a much better description, and this is particularly enhanced by the polarization functions.

This accounts for the structure of the *correlation-consistent* family of basis sets mentioned in section 2.6. For a nitrogen atom, for example, the cc-pVQZ quadruple-zeta basis set comprises five *s* basis functions (one for the 1*s* core, and four for the valence 2*s* atomic orbital), four *p* functions, but also three *d* functions, two *f* functions, and one *g* function. The *f* and *g* functions do not hugely improve the description of the molecular orbitals of molecules containing nitrogen atoms, and they do not make a major change to the Hartree–Fock energy. Instead, it is found that they much improve the description of the many-electron wavefunction and hence of the energy when using correlated methods. Significant changes in total energy and relative energy are found when improving the basis set in a correlated calculation, even when the basis set is already much larger than the triple-zeta and polarization standard generally required to obtain reliable Hartree–Fock results. This can be summarized in a diagram called a Pople diagram (named after John Pople, a British quantum chemist). Accurate results require *both* an accurate correlated method *and* a large basis set. Carrying out calculations with a sophisticated technique such as CCSD(T) but a small basis set such as 6-31G, or a very large basis set but a low level of theory, will *not* guarantee good results and is often a poor strategy.

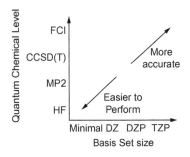

The convergence towards the correlation energy that would be obtained with an infinitely large basis set is unfortunately rather slow. This can be remedied in two ways. One is to attempt to *extrapolate* the total energy from values obtained with several basis sets. When using the correlation-consistent cc-pVXZ family of basis sets (X = D, T, Q, ...), quite smooth convergence of the correlation energy is obtained, with the deviation between the correlation energy obtained with a particular member of the set and the limiting value going as $1/x^3$, where x is the cardinal number of the basis (D: 2, T: 3, Q: 4, and so on). Knowing this, it is possible to estimate the energy that would be obtained at the infinite basis set limit based on calculations with two successive members of the family.

Another option is to include in the correlated method not only excited determinants in which the Hartree–Fock vacant orbitals are populated, but also additional functions that are constructed in such a way that they have an explicit dependence on the electron–electron distance. These *explicitly correlated* (or -F12, denoting their inclusion of a dependence on the distance between two electrons, 1 and 2) methods also provide a correction to the Hartree–Fock energy. They are complicated to formulate, and to incorporate in a program, so have only recently become available. Also, they do not modify the final result that a method gives, assuming that an infinitely large basis set is used. However, they do yield correlated energies *for a given finite basis set* that are much closer to the infinite basis set limit than a normal calculation does, without the computational challenge being much increased.

Correlation effects also impact significantly on calculated *relative* energies. Relative energies are important in quantum chemistry, because they relate to things like bond energies, reaction energies, and the like. State of the art methods such as CCSD(T)-F12 are able to predict relative energies for gas-phase reactions to within better than 5 kJ mol^{-1} for many small and medium-sized systems. New efficient implementations also allow application to larger molecules.

As already discussed, correlation energies are always large (except for one-electron systems), so much so that one might think that Hartree–Fock energies are simply not accurate enough to make any chemical predictions. This is not quite correct, however,

The effect of increasing the basis set

The table of total energies shown here illustrates the effect of increasing the basis set for the energy of the nitrogen molecule (at its experimental bond length of 1.09768 Å). The Hartree–Fock energy E_{HF} becomes progressively more negative upon moving from cc-pVDZ to cc-pV5Z. This is partly due to describing the valence orbitals better in the ways suggested in section 2.6, and partly due to an improved description of the $1s$ orbital. The polarization functions only contribute to the improvement of the valence orbitals, and their impact can be measured e.g. by considering the difference in HF energy between the full cc-pVQZ basis (including 3 d, 2 f, and 1 g function on each N atom) and a truncated cc-pVQZ basis with only s, p, and d functions.

The CCSD energy, and the correlation energy, also decrease with larger basis sets. Larger basis sets with many polarization functions much improve the description of electron correlation—E_{corr} is almost 0.1 hartree or over 260 kJ mol^{-1} more negative with cc-pV5Z than with cc-pVDZ. Also, while omitting the f and g functions from cc-pVQZ made only a relatively modest change to the HF energy, the impact on the CCSD energy is much larger. Finally, it can be noted that the cc-pV5Z energy is still some distance from the estimated infinite basis set limit, based on extrapolation from cc-pVTZ and cc-pVQZ (denoted here cc-pV[T,Q]Z). The energy obtained with cc-pVTZ, but with F12 explicit treatment of correlation is almost as good as that obtained with cc-pV5Z, but the computational effort is much smaller.

Basis	E_{HF}	E_{CCSD}	E_{corr}
cc-pVDZ	−108.95413	−109.26339	−0.30926
cc-pVTZ	−108.98347	−109.35536	−0.37188
cc-pVQZ	−108.99109	−109.38421	−0.39312
cc-pVQZ (spd)	−108.98756	−109.34447	−0.35691
cc-pV5Z	−108.99277	−109.39339	−0.40062
cc-pV[T,Q]Z	−108.99109	−109.39970	−0.40862
cc-pVTZ(F12)	−108.98914	−109.38736	−0.39822

because the correlation energy tends to cancel out in part when comparing energies of different species with the same atoms in different arrangements. The way in which this error cancellation occurs is somewhat predictable, based on some simple rules.

The first rule is that if two structures are *isodesmic*, i.e. they involve the same atoms, with the same number of chemical bonds of the same type, then their correlation energies will be very similar. For example, $CH_4 + CH_3–CH_2–CH_3$ and $2\ CH_3–CH_3$ are isodesmic structures, since both contain four carbons and twelve hydrogens, with two C–C bonds and twelve C–H bonds. The hypothetical chemical reaction $CH_4 + CH_3–CH_2–CH_3 \rightarrow 2\ CH_3–CH_3$ is referred to as an isodesmic reaction. What is meant by the 'same' type of chemical bonds is to some extent a matter of definition. For example cyclopropane + *n*-butane $\rightarrow$ cyclopentane + ethane can be considered to be an isodesmic reaction (6 C–C bonds and 16 C–H bonds in both reactants and products), but one might also consider that the high strain in the cyclopropane C–C bonds makes it preferable to count them separately.

A second rule is that if a hypothetical chemical transformation is *isogyric*, meaning that reactants and products have the same number of unpaired electrons, then even if they are not isodesmic, the correlation effect on the reaction energy will be small, albeit not as small as if the reaction was isodesmic. For example, the reaction $F + CH_4 \rightarrow HF + CH_3$ involves one unpaired electron in both reactants and products, and hence is isogyric.

An even more general rule is that the correlation energy for a system with n doubly occupied molecular orbitals is roughly equal to a constant multiplied by n. Unpaired electrons make a much smaller contribution to the correlation energy. Hence we expect the correlation energy in cyclopentane, with 15 valence doubly occupied orbitals, to be roughly 5/3 of that in cyclopropane, with 9 valence doubly occupied orbitals (core orbitals are typically not included in correlation treatment, as discussed above). This is a very loose rule of thumb, but it is helpful for assessing the likely magnitude of correlation effects.

3.7 Multi-reference methods

In many cases, especially for molecules near their equilibrium structure, correlation effects lead to a small correction to HF, with the HF wavefunction providing a reasonable starting point. However, upon stretching bonds away from equilibrium, correlation effects can become much larger, such that the HF wavefunction is no longer even qualitatively correct. A typical example is the hydrogen molecule, H_2. Neglecting spin and antisymmetrization, the Hartree–Fock wavefunction can be written as shown in equation (3.8):

$$\Psi_{HF}(r_1, r_2) = \psi_\sigma(r_1) \times \psi_\sigma(r_2) \tag{3.8}$$

Where the σ molecular orbital ψ_σ is an in-phase combination of the $1s$ atomic orbitals on atom A and atom B, ϕ_A and ϕ_B, equation (3.9) (N is the normalization constant)

$$\psi_\sigma(r) = N(\phi_A(r) + \phi_B(r)) \tag{3.9}$$

The Hartree–Fock wavefunction can be rewritten by inserting equation (3.9) in equation (3.8) and expanding, to yield:

$$\Psi_{HF}(r_1, r_2) = N^2[\phi_A(r_1)\phi_A(r_2) + \phi_A(r_1)\phi_B(r_2) + \phi_B(r_1)\phi_A(r_2) + \phi_B(r_1)\phi_B(r_2)] \tag{3.10}$$

On inspecting this wavefunction, it can be seen to yield a qualitatively incorrect description of H_2 at very large bond length. Physically, one should simply have two separate hydrogen atoms, so the wavefunction should predict 100% probability for situations where r_1 is close to one nucleus, and r_2 is close to the other. Instead, Ψ_{HF} is a mixture of four terms, with equal amplitudes (hence equal probabilities upon squaring the wavefunction). The second term is of the expected form, large when electron 1 is near nucleus A, and electron 2 is near nucleus B. The first term, however, is large when *both* electrons are near nucleus A. This is equivalent to saying that the arrangement H^- .... H^+ is likely to occur. The fourth term is likewise unphysical. Even though Figure 3.2 was calculated at the equilibrium bond length, the same general appearance would be obtained at large bond length, with four peaks of equal magnitude corresponding to the four terms in equation (3.10). As a result, HF yields a very incorrect energy for H_2 at large bond length. This energy should be –1 hartree (the energy of a hydrogen atom is –0.5 hartree), and should no longer vary with r_{HH} beyond

A deeper look at correlation energies

To illustrate the rules stated in the main text, the table below contains calculated HF and correlated energies for two reactions that are neither isodesmic nor isogyric, three that are isogyric but not isodesmic, and two that are isodesmic. The energies of products are given relative to that of reactants, in kJ mol^{-1}. As well as the Hartree–Fock result, values are given based on calculations with a simple correlated method, MP2, and a highly sophisticated one, CCSD(T)-F12.

What this means is that a calculation is done at the indicated level for all the products using a structure similar to their experimental structure, and the resulting total energies are added together. This gives the total energy for the products, considered infinitely far apart (note that we use here the fact that all methods used are size-consistent). Then the same is done for the reactants. Finally, the two energies are compared, and the difference is expressed in kJ mol^{-1} (1 hartree = 2625.5 kJ mol^{-1}). The HF and MP2 calculations use the 6-31G(d) basis set, with unrestricted HF for the radicals. The CCSD(T)-F12 calculations use the cc-pVTZ-F12 basis set, except for the case of the last entry, where cc-pVDZ-F12 was used. In all cases, the optimum geometry from MP2/6-31G(d) and a correction for zero-point energy at the same level were used (see Chapter 5 for an explanation of these terms). The experimental data was taken from the NIST database.

For the first two reactions, which correspond to bond-breaking, the correlation

Reaction	ΔE (HF)	ΔE (MP2)	ΔE (CCSD(T)-F12)	$\Delta H^0_{298,exp}$
$N_2 \rightarrow 2\,N$	418.3	883.0	928.0	945.4
$CH_3OH \rightarrow CH_3 + OH$	209.5	374.9	375.0	386.2
$NH_3 + OH \rightarrow NH_2 + H_2O$	−10.9	−39.9	−50.1	−44.5
$F + CH_4 \rightarrow HF + CH_3$	−20.5	−99.0	−127.9	−132.1
$C_2H_4 + H_2 \rightarrow C_2H_6$	−146.9	−133.3	−126.4	−136.5
$CH_3OH + C_2H_6 \rightarrow CH_4 + C_2H_5OH$	−21.2	−26.1	−24.6	−23.4
$c\text{-}C_3H_6 + n\text{-}C_4H_{10} \rightarrow c\text{-}C_5H_{10} + C_2H_6$	−87.4	−90.1	−88.7	−88.1

effect is very strong, with the starting molecule being much more stable relative to the fragments with the correlated method than with Hartree–Fock. The effect is biggest for N_2, where three bonding pairs split to give a total of six unpaired electrons in the two nitrogen atoms formed upon bond-breaking. The HF bond energy in this case is over 500 kJ mol^{-1} smaller than experiment. Note that the experimental data given relates to gas-phase enthalpies at room temperature, and thereby includes a number of effects not treated in even the best calculation in the table. Hence perfect agreement is not expected, but one can see that CCSD(T)-F12 is always very close to experiment.

For the three isogyric reactions, HF is closer to CCSD(T)-F12, though the extent of the difference is variable. For the hydrogenation reaction $C_2H_4 + H_2$, Hartree–Fock predicts the reaction energy accurately, whereas for the $F + CH_4$ reaction, it does not. This is because the rule of thumb whereby electron correlation effects only depend on the number of pairs of valence electrons is just that—a rule of thumb. The correlation energy in each molecule is determined by many factors. Finally, for the two isodesmic reactions, the agreement between Hartree–Fock, coupled-cluster, and experiment is excellent. Even for the case involving strained cyclopropane, good results are obtained.

≈ 3 Å. Instead, HF yields a higher energy (e.g. for $r_{HH} = 10$ Å, $E = -0.74$ hartree) that continues to climb (Figure 3.4). The correlation energy has become very large.

As in equation (3.1), the model can be improved by using a more complicated wavefunction, with a contribution from multiple Slater determinants. In fact, a significant improvement is obtained simply by mixing in just one doubly excited determinant, where both electrons have moved to the σ^* anti-bonding molecular orbital:

$$\Psi_{CID}(r_1,r_2) = c_0\{\psi_\sigma(r_1) \times \psi_\sigma(r_2)\} + c_D\{\psi_{\sigma^*}(r_1) \times \psi_{\sigma^*}(r_2)\} \tag{3.11}$$

This wavefunction has been labelled 'CID' in equation (3.11) and in Figure 3.4, as it is a very simple CI-doubles wavefunction, with only *one* doubly excited determinant included. Compared to normal CI, it is noteworthy that the mixing coefficient c_D is not small: upon optimization of this coefficient using the variational principle, at large r, it becomes equal in magnitude (and opposite in sign) to c_0. The resulting approximate wavefunction Ψ_{CID} yields the correct physical trend for large r, with the energy flattening out by $r \approx 3$ Å. However, the energy at large r is well above the correct value of -1 hartree. This is because the shapes of the σ (and σ^*) molecular orbitals have been determined by the Hartree–Fock calculation, so as to minimize the energy of the Hartree–Fock wavefunction Ψ_{HF}. The resulting orbitals are however not optimal for the Ψ_{CID} wavefunction.

To obtain a better result, one can also optimize the shape of the orbitals. In other words, one can use a variational approach in which both the coefficients defining the contributions of the different Slater determinants, and the coefficients of equation (2.11) that define the contribution of the atomic orbitals to the molecular orbitals are simultaneously varied. This yields a family of methods known as 'multi-configuration self-consistent field' (or MCSCF) methods. The most commonly used such method is known as the *complete active space* self-consistent field (or CASSCF) method, in which all Slater determinants that can be obtained by moving a certain number of electrons

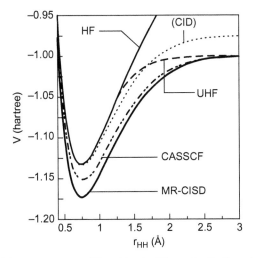

Figure 3.4 Potential energy curves at different levels of theory for the H_2 molecule.

between a certain number of orbitals are included. The set of orbitals treated specially in this way is called the *active space*. In Figure 3.4, the active space used was simply the σ and σ^* orbitals, which in principle would allow for four determinants. For symmetry reasons, however, only the two configurations shown in equation (3.11) can contribute. Ψ_{CASSCF} yields the correct energy and slope at large r.

As can be seen in Figure 3.4, the CASSCF energy is not only lower than E_{HF} for large r, but also for $r = 0.75$ Å, the equilibrium bond length. This is because the correlation effects described in sections 3.2 to 3.5 occur in H_2, and are partly described by mixing in the σ^{*2} configuration. However, to obtain truly accurate results for this molecule, additional excited configurations must also be included, by taking into account many different orbitals that are vacant in the HF wavefunction, not just the σ^* orbital. The shape of these orbitals does not, however, strongly need optimizing, so a two-step procedure can be used, in which the two key orbitals are first optimized in a CASSCF procedure, then additional excited determinants are considered while including only the coefficients of equation (3.1) in the variational treatment. This is the result labelled MR-CISD in Figure 3.4, and it yields exact results (within the limitations of the basis set).

The final curve in Figure 3.4 is labelled UHF, and is indeed a UHF calculation. As can be seen, it converges to the correct energy of −1 for large r, whereas at small r, it overlaps with the standard HF curve. Standard HF fails for H_2 at large r as explained by equation (3.10). The UHF calculation shown here involves 'breaking the symmetry' of the problem: the two spatial orbitals used in the calculation, ψ_α and ψ_β, can optimize in such a way that they no longer have the symmetry of the nuclear framework: at large r, ψ_α will optimize to become the $1s$ orbital on atom A, while ψ_β will become the $1s$ orbital on atom B (or vice versa). Such 'broken-symmetry' HF wavefunctions suffer from some disadvantages but have the benefit of returning at least qualitatively correct energies and are frequently used in studies of systems with multiple unpaired electrons, in combination with correlated methods or density functional theory.

Returning to the CASSCF method, it can be applied for much larger systems than H_2, with the same general approach of including *all* possible electron distributions within a limited number of orbitals, the active space. By including all possible determinants, the method is similar to full CI, with the additional complication of requiring optimization of the orbital coefficients. The method is nevertheless computationally tractable provided only a limited active space is used, typically of 16 or fewer orbitals. The approach is shown in schematic form in Figure 3.5, with a small active space of three orbitals, containing four electrons. In principle, the electrons can be spread in these orbitals in nine different ways (only three are shown). Note that orbitals lower in energy than the active space are always doubly occupied, and those above it are always vacant.

The MCSCF method is essentially a procedure for treating electron correlation, as are the other methods that include multiple determinants in some way, such as CISD, MP2, and CCSD discussed in previous sections. What, then, is the difference? As mentioned above in the case of H_2, MCSCF approaches are needed where some 'excited' determinants contribute very significantly to the wavefunction: for H_2 at large bond length, to get the qualitatively correct behaviour, the coefficient c_D for the excited determinant in equation (3.11) must be equal in magnitude (though

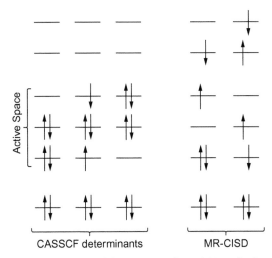

Figure 3.5 Schematic representation of the CASSCF and MR-CISD methods.

opposite in sign) to the coefficient for the ground-state or Hartree–Fock determinant, c_0. As will be seen below, correlated methods based on an admixture of small amounts of many excited determinants into the Hartree–Fock wavefunction (such as MP2 and CCSD), do not yield good results in such cases. This different scope of applicability has suggested that electron correlation can be of two different types: 'normal' correlation, requiring a large number of small albeit important corrections to the wavefunction, with the HF wavefunction being used as a reasonable starting point, and 'special' correlation, where HF is qualitatively wrong and several determinants are needed in order to get even a roughly reasonable wavefunction. 'Normal' correlation is usually called **dynamic** correlation, and 'special' correlation is usually called **static** correlation, or '**non-dynamic**' (in other contexts, it is also called 'strong' correlation).

While many molecular systems do not display static correlation, dynamic correlation occurs nearly universally, including in systems with static correlation. Because of the limited number of active orbitals in CASSCF, it does not actually do a particularly good job of treating dynamic correlation. This can be seen in Figure 3.4 for H_2, where, near the equilibrium bond length, the wavefunction including higher excitation, labelled MR-CISD, yields an energy that is lower than CASSCF. In fact, H_2 is almost an exception, since at large r, CASSCF yields the exact result (the dissociated H atoms each only have one electron, hence there is no more correlation *within* the atoms, only between them). A more typical case is shown in Figure 3.6, for N_2, where HF, CASSCF, and MR-CISD calculations are shown. The active space used includes eight orbitals, namely two σ and two σ^* orbitals, and two π and two π^* orbitals, that can be constructed by mixing the $2s$ and $2p$ orbitals on the two atoms. The MR-CISD method has been mentioned already but not well defined. The acronym refers to *multi-reference configuration interaction with single and double excitations*. It is a multi-configuration version of CISD, which allows the dynamic correlation effects to be

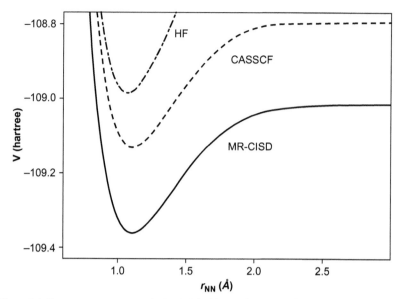

Figure 3.6 Energy curves computed using HF, CASSCF, and MR-CISD for dinitrogen using the cc-pVTZ basis set.

treated as well as the static correlation. An MR-CISD calculation is performed in two steps: first, a CASSCF calculation is carried out. Then, as well as all the CASSCF Slater determinants, additional determinants in which one or two electrons from the active space are moved into higher-lying orbitals are included (see Figure 3.5), and their coefficients are variationally optimized. As well as using a multi-configuration version of CI, it is also possible to combine perturbation theory with multi-reference methods. This can be done in several ways, leading to different methods such as CASPT2 (*complete active space with perturbation theory at second order*). Unlike for the unusual case of H_2, in Figure 3.6, MR-CISD leads to a lower energy than CASSCF in the case of dinitrogen at all bond lengths, since dynamic correlation between the electrons on each nitrogen atom plays a role even upon dissociation.

The nitrogen molecule illustrates the by now familiar shortcomings of the Hartree–Fock method upon dissociating bonds. However, it is also useful to illustrate the problems encountered by the correlated methods based on the Hartree–Fock single reference starting point covered in previous sections. Figure 3.7 shows potential energy curves for N_2 computed with MP2, CCSD, and CCSD(T), which all perform fairly well near the equilibrium geometry, but which all undergo more or less catastrophic failure at large r as the Hartree–Fock wavefunction starts to become an unrealistic starting point for describing correlation.

CASSCF, MR-CISD, and CASPT2 methods are frequently used where HF fails in qualitative terms, e.g. for stretched molecules, for excited states, and for some molecules where HF is inadequate, for various reasons, even at the equilibrium geometry. This typically occurs where a molecule cannot be well described using a single Lewis (dot and cross) structure, or a combination of Lewis structures in resonance with one another.

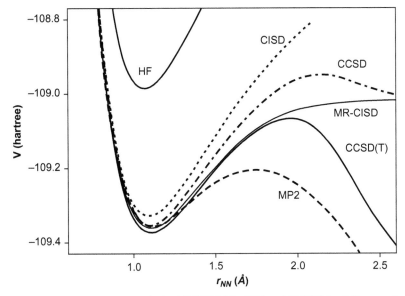

Figure 3.7 Energy curves computed using HF, CISD, MP2, CCSD, and CCSD(T) for dinitrogen using the cc-pVTZ basis set. The near-exact MR-CISD curve from Figure 3.6 is shown for reference.

3.8 Density functional theory

The previous sections describe methods to solve the electronic Schrödinger equation, equation (2.2), through construction of approximate many-electron wavefunctions of the general form shown in equation (3.1). A rather different approach exists in which the central object is no longer the wavefunction, but instead the electron density distribution, $\rho(r)$. For a many-electron system, the density is a much simpler function than the wavefunction. Nevertheless, the physicists Pierre Hohenberg and Walter Kohn showed that it is possible, in principle, to calculate the energy of the electronic ground state of a system as a functional (a generalization of a function) of its density, $E[\rho(r)]$. They also showed that there is a variational principle for the density $\rho(r)$: any approximate density ρ' would return a higher energy than the correct ground state density, $E[\rho'] > E[\rho]$, would.

These statements, known as the Hohenberg–Kohn theorems, suggest that a density functional theory (DFT) could be a fruitful approach to use in quantum chemistry, but they do not address exactly *how* to calculate the energy. In other words, they do not provide a mathematical expression for the energy functional, $E[\rho]$. Many years after the theorems were proven, it remains true that the *exact* expression for the density functional is not known. Instead, *approximate* functionals have been developed. In early work, attempts were made to formulate functionals in which only the density itself appeared as an input. However, such methods struggle to account accurately for the kinetic energy part of the electronic energy, and following work by physicists Walter Kohn and Lu Jeu Sham, an alternative approach was used in which the density was written (equation (3.12)) as the sum of densities from a set of molecular orbitals (also known as Kohn–Sham orbitals):

$$\rho(r) = \sum_{i}^{n_{elec}} \psi_i^2(r) \tag{3.12}$$

Equation (3.12) leads to a straightforward way to compute a good approximation to the electronic kinetic energy. The overall energy expression can then be written as shown in equation (3.13).

$$E[\rho(r)] = \tfrac{1}{2}\iint \frac{\rho(r_1)\rho(r_2)}{r_{12}}dr_1\,dr_2 + T_S[\rho(r)] - \sum_J\left(\int \frac{\rho(r)Z_J}{r_J}dr\right) + V_{XC}[\rho(r)] \tag{3.13}$$

On the right-hand side of equation (3.13), the first term gives the Coulombic repulsion energy between the electrons making up the density, which can be computed straightforwardly. The second term gives the kinetic energy associated with the orbitals of equation (3.12), which is a good approximation to the exact kinetic energy. The third term gives the Coulombic energy associated with the interaction between the electron density and the nuclei (Z_J is the nuclear charge of the Jth nucleus), and this too is straightforward to calculate. However, the fourth term, called the exchange-correlation energy functional, is less straightforward: it comprises all the remaining unknown parts of the exact density functional. Based on known expressions for this term for some simple situations (such as a 'gas' of electrons of uniform density), it is possible to suggest physically reasonable (but approximate) functional forms for this term. It is also possible to fit parameters within flexible possible functional forms for this term, so as to reproduce some known theoretical or experimental results. Through this combination of theory-based reasoning and empirical fitting, quite accurate exchange-correlation functional forms have been developed. In many cases, separate functional expressions are obtained to calculate the exchange and correlation parts of the exchange-correlation energy.

In the framework of equation (3.13), to find the shapes of the molecular orbitals, one needs to solve a set of equations very similar in form to the Fock equations (2.10), the Kohn–Sham equations. Like the Fock equations, the Kohn–Sham equations contain terms that depend on the orbitals, so they need to be solved iteratively. In fact, the procedure for running a Kohn–Sham DFT calculation is remarkably similar to that used for HF: choose a system, a structure, a basis set, choose initial (or 'guess') orbitals, generate the Kohn–Sham equations and solve them, and cycle until convergence. The orbitals also have a similar shape to those obtained in HF, and can likewise be analysed to probe aspects of chemical bonding. Almost all of the text in Chapter 2 describing procedures for improving the convergence and efficiency of Hartree–Fock calculations is also relevant to Kohn–Sham DFT calculations and will not be repeated here. It should be noted that calculations typically require a basis set, which must satisfy the same requirements as that used in HF calculations, i.e. it should be multiple-zeta and allow for polarization.

The difference lies in the fact that the answer is an (approximate) DFT energy, not the HF energy. Many different exchange-correlation functionals are commonly used. The simplest of these compute the exchange-correlation energy by integrating over the whole volume of the molecule, and including energy terms that depend only on the electron density in each infinitesimal volume element. These are called *local* exchange-correlation functionals and are seldom used in chemical applications as they are barely more accurate than HF (though they do yield useful accuracy for some types of solid-state systems, considering their simplicity).

The next higher level of sophistication comprises exchange-correlation functionals in which for each infinitesimal volume element within a molecule, not only the electron density but also its gradient is taken into account. This allows for the effect that variations in density have on the local contribution to the energy. So-called gradient-corrected functionals, or functionals using the 'generalized gradient approximation' (GGA) are very commonly used. Many different GGA functionals have been suggested, and it can seem at first very challenging to learn to recognize them and the acronyms by which they are known. Different GGA functionals may yield better results for different classes of problems. Two commonly used GGA functionals are known as 'BP86' (developed by theoretical chemists Axel Becke and John Perdew, with one part of the work performed in 1986) and 'PBE' (developed by John Perdew, Kieron Burke, and Matthias Ernzerhof in 1996).

Going beyond the gradient, the second derivative or Hessian of the electron density can also be used, leading to so-called meta-GGA functionals. The second derivative of the density is related to the kinetic energy, so such functionals are also sometimes known as kinetic energy density functionals. One popular family of such functionals has been developed in the group of theoretical chemist Donald Truhlar in Minnesota, with functionals having names such as M06 (developed in 2006).

Another popular family of functionals includes some terms from conventional *ab initio* methods in the energy expression. The most common term to include is that describing *exchange* interactions between electrons, based on the integrals K_{ij} from equation (2.8) and the Kohn–Sham orbitals. Functionals including this type of exchange energy term, loosely known as 'Hartree–Fock' or 'exact' exchange, are known as *hybrid* density functionals. One of the most widely used functionals is based on an approach proposed by the Canadian theoretical chemist Axel Becke, and combines an exchange energy expression developed by him, the HF exchange energy expression, and a correlation energy expression due to Chengteh Lee, Weitao Yang, and Robert Parr, with three parameters governing the nature of the mix. It is known by the acronym B3LYP. The PBE GGA functional has a related hybrid form known as PBE0.

DFT accounts for correlation effects, and hence is much more accurate than HF. Studies on the energies required to split a set of small molecules into atoms (atomization energies) show that many DFT functionals have errors of only 10–20 kJ mol^{-1}, compared to a hundred kJ mol^{-1} or more for HF. The accuracy is similar to or better than that of MP2, and approaches that of CCSD(T). Such computations of mean errors on atomization play an important role in developing and assessing new DFT functionals.

For large molecules, it has recently been discovered that many functionals lead to incorrect energies and structures in some cases, because one particular sort of correlation effect is treated very poorly: London dispersion forces, the interaction between a momentary dipole arising as a random fluctuation in one part of a system with the momentary dipole that it induces in another part. These interactions are negligible in the small molecules used in the atomization energy datasets often used to parameterize functionals, but become larger in bigger molecular systems, and play a large role in intermolecular interactions. Interaction between two argon atoms is a simple example of a case where the only attractive component of the energy is due to dispersion. Potential energy curves for two argon atoms are shown in Figure 3.8. Hartree–Fock calculations neglect dispersion interactions completely and so show no attraction between the two non-polar atoms (although basis set superposition error introduces a very shallow minimum in the potential energy curve at large Ar–Ar distances). The correlated methods such as MP2 or CCSD(T) provide a good description. Different

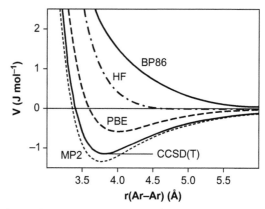

Figure 3.8 Ar–Ar interaction energy curves computed using HF, MP2, and CCSD(T) for reference, and BP86 and PBE for examples of DFT methods. In all cases the aug-cc-pVTZ basis was used.

DFT functionals account for dispersion to different extents, as is shown in Figure 3.8 for the two examples of BP86 and PBE. BP86 predicts that the interaction between these two closed-shell rare gas atoms is even more repulsive than does HF, whereas PBE is qualitatively correct in that there is a minimum, but it has the wrong interatomic distance and interaction energy compared to CCSD(T). Modern functionals are often parameterized including larger datasets for fitting, including some intramolecular interaction energies as well as atomization energies, so as to ensure a better description of dispersion. Also, some DFT functionals can be modified to include separate semi-empirical dispersion corrections.

3.9 Semiempirical methods

The methods described in Chapter 2 and in sections 3.2 to 3.7 are called *ab initio* ('from the beginning', in Latin) methods—they take as input only the Schrödinger equation, and make no use of results from experiment (except in the sense that experimental results are used to calibrate how well the different methods work). Density functional theory is also an *ab initio* method in some sense, though in some cases the density functional expressions are fitted to some experimental datasets as well as to theoretical requirements. Yet other quantum chemical methods exist that make explicit and systematic use of experimental (empirical) data in their elaboration, while maintaining the quantum mechanical framework, and these are referred to as *semiempirical* methods.

One important family of such methods is based on Hartree–Fock theory. These methods use a wavefunction of the form of equation (2.5) based on molecular orbitals, but instead of using the normal energy expression of equation (2.7), some of the integrals of the type shown in equations (2.12) and (2.13) are set to zero, especially many of those involving basis functions situated on different atoms. Other integrals are given adjusted values, chosen to improve agreement between calculated energies and some target value. Originally, the energy data used for fitting came from experiment, hence the reference to these methods as *semiempirical*, though nowadays, accurate calculations may be used instead. Three heavily used semiempirical methods are referred to through their acronyms—AM1 (from Austin Model 1, developed by the British/American theoretical chemist Michael Dewar in Austin, Texas), PM3 (parameterized method three), and PM6.

Semiempirical methods based on density functional theory also exist, such as the DFT-based tight-binding approach.

Semiempirical methods are much less computationally demanding than *ab initio* methods, even than HF, yet if they are well parameterized, they can be quite accurate—certainly they are typically much more accurate than HF, though usually also less accurate than good DFT or correlated methods. They can therefore be used where calculations on large systems (or many repeated calculations) are needed.

3.10 Solids and periodic models

Until now in this chapter, it has been implied that quantum chemical calculations are carried out on isolated molecular systems, considered to be present as a collection of particles surrounded by vacuum. In chemistry, the systems of interest are often larger assemblies of molecules, or condensed phase systems such as liquids or solids. To model such systems, one possibility is to construct larger and larger models, consisting of multiple interacting molecules, and to treat them with the methods described so far in this chapter. A second possibility is to treat them using hybrid methods, whereby some core region is described using quantum chemical techniques, and the environment is treated in a different, simpler way. Such approaches are discussed in Chapter 8. There is a third approach, which is well suited to crystalline solids, and which consists of treating a *periodically repeating model*. The details of this approach—that is mainly used in conjunction with DFT—go beyond the scope of this book, but some elementary principles are introduced here.

The general idea is to transform an impossible problem—modelling of an infinitely large system—into a tractable one, by exploiting the translational symmetry. In many cases, calculations with the resulting methods can be carried out with a computational effort roughly similar to that needed to carry out a similar calculation on the periodically repeated unit.

Consider a simple example: a one-dimensional linear 'crystal' of helium atoms, with a He–He distance of L. The first approach to calculate properties of this system would be to build a model of the full system judged to be large enough to represent the properties of the whole, and containing a few atoms—one, two, or more atoms—and to carry out standard quantum calculations on it. Making two models of different size could be done to check that indeed the desired property is well represented. In the second approach, some 'core' (e.g. He_3) would be treated quantum chemically, with some additional energy terms describing interaction with the neighbouring atoms.

In the third approach, the topic of this section, a calculation is carried out on the infinitely long, periodically repeating system, with helium atoms positioned at $x = 0$, $y = 0$, and $z = iL$, where i is any integer. Periodic calculations of this type can also be performed with two periodically repeating dimensions, or three. In the discussion here, only the one-dimensional case will be considered, for reasons of simplicity.

For the periodic system, it is clear that the electron density must be translationally symmetric: $\rho(z) = \rho(z+L)$. Within DFT, this also requires that the *square* of each of the orbitals be translationally symmetric, so that $\psi^2(z) = \psi^2(z+L)$, or, more generally if the orbitals are allowed to adopt complex values, that $\psi\psi^*$ be translationally symmetric in this way. The easiest way for this to be achieved is for $\psi(z)$ to be equal to $\psi(z+L)$ for all orbitals, but one can also have $\psi(z) = -\psi(z+L)$, or, more generally, $\psi(z) = e^{i\theta}\psi(z+L)$, where θ is an arbitrary angle and i is the square root of -1. This condition is automatically satisfied provided that the molecular orbital is written as a product of a spatial part $u_n(z)$, that is strictly symmetric with respect to translation, and a complex number term e^{ikz}:

$$\psi_{n,k}(z) = e^{ikz} \times u_n(z) \tag{3.14}$$

In this equation, k is a constant that can adopt a range of possible values, and each of these values corresponds to a different orbital of the whole system. In principle, one needs to consider all possible values of k to describe all the possible ways in which ψ can vary across the crystal. However, it can be shown that only values of k between 0 and π/L (which is the value such that $e^{ikL} = -1$) need to be considered. In practice, a DFT calculation on a periodic system requires consideration of a finite set of values of k (or 'k points') in each direction (for a three-dimensional system, k is a three-dimensional vector), to range between 0 and π/L. Where a molecular calculation yields a certain number of molecular orbitals, each with one energy, a periodic calculation will yield a certain number of *families* (or bands) of orbitals. Each band will have the same fully symmetric part u_n, but different orbitals within this band will have different values of k and different energies. In the helium example, if the repeating unit is chosen to be just one helium atom, then there will be only one occupied band, associated with multiple values of k. In general, there will be multiple occupied bands (multiple different u_n), each associated with multiple values of k.

Periodic calculations use a greater diversity of basis sets than molecular calculations, for which the Gaussian basis sets as described in section 2.6 are almost universally used. In periodic calculations, Gaussian functions can also be used, but so can so-called *plane wave* basis sets, or mixtures of Gaussian functions and plane waves. Note that if the repeating unit in a periodic system is considered to be a box, with a molecule in its centre surrounded by vacuum, then periodic quantum chemistry codes can be used to model isolated molecules.

3.11 Molecular properties

Much of the discussion and of the examples given above have focused on the calculation of molecular *energies*. As will be discussed at length in Chapters 5 and 6, the energy is indeed a very central quantity in physical chemistry. Nevertheless, the wavefunctions and densities computed with quantum chemical methods can also be used to predict other molecular properties.

The overall electrostatic properties of molecules, such as their *dipole moment* and higher-order multipoles, or the electrostatic potential at points around the molecule, are one example. These are obtained directly from the calculated wavefunction. As for energies, each method has a characteristic level of accuracy for such quantities and this should be kept in mind when comparing results to experiment. For example, at the HF level of theory, the dipole moment of hydrogen fluoride is computed to be 2.02 D, versus 1.86 D from experiment ('D' here refers to the traditional unit for dipole moments, the debye).

Atomic charges are not directly observable in experiments, though they are often invoked in discussions of bonding. They are linked to the electrostatic properties. Approximate charges can be computed in many different ways. One frequently used type of approach divides the overall electron population of an orbital between the atoms in the molecule. The most common method of this type was suggested by the American physical chemist Robert Mulliken, and leads to so-called 'Mulliken charges'. Another type of approach seeks to find the atomic charges that best reproduce the electrostatic potential field around the molecule. The CHELPG approach is one such frequently used method. Once again, since atomic charges cannot in principle be observed, there is no single way to define them in quantum mechanics, so it cannot be expected that they can be calculated in a single 'best' way.

All sorts of spectroscopic properties, including those relating to nuclear magnetic resonance (NMR) spectroscopy, can also be calculated using quantum chemistry. *Infrared* line positions and intensities are related to vibrational frequencies and will be discussed in Chapter 5.

One important property of a molecular system is the nature of its electronically excited states. Most of the methods discussed in this chapter, at least in their standard form, are only applicable to the electronic ground state of a system. However, many methods are also available for calculating electronically excited states. The basic idea is to use the excited electronic configurations shown in Figure 3.1 and in equation (3.1). Configuration interaction and multi-configuration self-consistent field (MCSCF or CASSCF) methods are particularly suited to such calculations. A modification of DFT called time-dependent DFT (or TDDFT) exists that can also be used for exploring excited states.

3.12 Further reading

- A priori calculation of molecular properties to chemical accuracy, T. Helgaker, T. A Ruden, P. Jørgensen, J. Olsen, and W. Klopper, *Journal of Physical Organic Chemistry*, 2004, **17**, 913–933. This review provides a relatively accessible description of the accuracy of quantum chemical methods for describing correlation effects.
- *Density-Functional Theory of Atoms and Molecules*, Robert G. Parr and Yang Weitao, Oxford University Press, Oxford, 1995. This advanced book describes the development and underlying theory of DFT methods.

3.13 Exercises

3.1 Carry out Hartree–Fock, MP2, and CCSD(T) calculations using a medium-sized basis set such as 6-31G(d) for the reactants and products of some simple chemical reactions similar to those shown in section 3.6. Select reactions that are isodesmic or isogyric or neither of these. Calculate the reaction energy. For this study, it will be helpful to first obtain reasonably accurate structures for each of the reactants and products—this can be done either by consulting a database of molecular structures (such as Pubchem) or by using geometry optimization, as discussed in Chapter 5. Use a database of experimental thermodynamic data, such as the online NIST Chemistry Webbook, to obtain experimental values for the reaction energy, and compare to your calculations.

3.2   Use density functional theory, together with the BP86 functional (or another GGA) to compute the energies for the species in exercise 3.1, and compare the accuracy of the relative energies as well as the time needed for the calculations with those in exercise 3.1.

3.3   Carry out CASSCF calculations for the hydrogen fluoride molecule for various values of the H–F distance. Use a medium-sized basis set such as 6-31G(d), and an active space in which six electrons are allowed to occupy the two π-symmetric lone pair orbitals, the σ bonding and σ^* antibonding orbital. Generate a plot of the energy and estimate the dissociation energy.

3.4   Use a DFT method such as BP86 together with a medium-sized basis set, such as 6-31G(d), to calculate the absolute chemical shift in nuclear magnetic resonance for the hydrogen nuclei in the benzene and methane molecules. Experimentally, the chemical shifts for the hydrogen nuclei in these two molecules are roughly 0.2 and 7.3 ppm, respectively.

3.14 Summary

- Due to its product form, the Hartree–Fock wavefunction neglects *correlations* in the positions of electrons, but correlated methods exist that can yield improved results.

- Particular noteworthy methods include *configuration interaction* methods such as CISD or FCI, the perturbation-based MP2, the coupled-cluster CCSD(T) approach, and the multi-reference CASSCF, MR-CISD, and CASPT2 methods.

- As well as *ab initio* methods, semiempirical methods and density functional theory are extremely useful methods for solving the Schrödinger equation.

- Quantum chemical methods yield energies that can be combined to obtain reaction energies, and these can be predicted very accurately using some methods.

- Quantum chemistry can also be used to characterize many molecular properties.

- As well as calculations on isolated molecules, computations of wavefunctions and energies for periodic crystalline systems are possible.

4 Molecular Mechanics Methods

4.1 Introduction

The previous two chapters described quantum chemical methods to calculate the potential energy for a given species and for a particular arrangement of the nuclei, $V(R)$. This is a powerful and very general way to obtain V, but it is also rather complicated, and hence time-consuming. It can also yield results that are not accurate, especially for the more simple approximate methods. In this chapter, an alternative, less general but less time-consuming method is introduced: molecular mechanics (MM). In this approach, a known chemical bonding pattern is assumed, and used to define preferred bond lengths and angles, and thereby an energy expression that takes into account distortions away from these ideal values. This is analogous to the way in which the mechanical properties of the balls and sticks in a traditional molecular model convey a mixture of rigidity and flexibility to the model. For a given bonding environment, the type of energy terms needed, and the numerical parameters within the energy expression, are transferable from one system to another. Hence general **forcefields** can be constructed with quite general applicability.

The chapter will describe how the energy terms and parameters are chosen, based on input from experiment and quantum chemistry. MM can be applied to large systems due to its efficiency, allowing calculations on liquids, solutions, and solids. This frequently makes use of periodically repeating models and the special measures needed to treat such models are also discussed. Finally, the type of software used for MM is discussed.

What is meant by 'forcefield'?

The word 'forcefield' is used a lot in the context of molecular mechanics methods, so it is important to know what it means. Like many frequently used concepts, it has a variety of meanings, two of which are important here. The first meaning is 'an equation used to calculate the energy of a specific model system, comprised of a collection of atoms or atom-sized particles, and to calculate the derivative of the energy, i.e. the forces acting on the atoms.' In this meaning, any system one chooses to study will be described by its *own* forcefield. Section 4.2 describes how such a forcefield is constructed. The second meaning is 'a set of generic rules for constructing a molecular mechanics energy expression for a member of a general class of model systems, including a definition of which energy terms should be used, and which parameter values should be used in the energy terms.' For example, one can talk about 'the TIP3P forcefield for liquid water' (it can be applied to many different sizes of models of liquid water, from just one or two molecules to millions of them), or 'the AMBER protein forcefield' (it can be applied to any polypeptide system).

4.2 MM forcefields

Consider a simple molecule, such as water. This has two O-H bonds, and it is known from experiment, as well as qualitative and quantitative molecular electronic structure theory, that these bonds have a preferred bent structure, with an H-O-H angle of approximately 100°. A simple energy expression that takes these features into account is:

$$V(r_A, r_B, \theta) = k_{stretch}(r_A - r_0)^2 + k_{stretch}(r_B - r_0)^2 + k_{bend}(\theta - \theta_0)^2 \qquad (4.1)$$

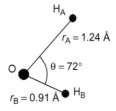

H_A
$r_A = 1.24\ \text{Å}$
O
$\theta = 72°$
$r_B = 0.91\ \text{Å}$ H_B

In equation (4.1), V is the overall potential energy of the system for given values of the two O-H distances r_A and r_B, and of the angle between the two bonds, θ. The values $k_{stretch}$ and k_{bend} are *force constants*—fixed-value numbers chosen to try to describe the potential energy changes for stretching and bending as well as possible. And r_0 and θ_0 are reference values for the stretch and bend, corresponding to the equilibrium structure of the molecule.

Consider a slightly more complicated molecule, methanol. Here the corresponding molecular mechanics expression will have many more terms. There will be five bond stretching terms, one each for the three C-H bonds, the one C-O bond, and the one O-H bond. There will be seven bending terms, corresponding to the three distinct H-C-H angles, the three distinct H-C-O angles, and the C-O-H angle. And there will be three terms of a new type, depending on the *dihedral* angle α (H-C-O-H). Various mathematical forms can be used for this term—one of the simplest is shown as equation (4.2):

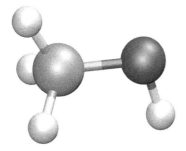

$$V(\alpha) = A_{dihedral}\{1 + \cos(n(\alpha - \alpha_0))\} \qquad (4.2)$$

where $A_{dihedral}$ is a constant defining the 'stiffness' of the system towards rotation around the central bond defining the dihedral angle, α_0 is the reference angle (for which the energy term adopts its largest value), and n is an integer that defines the *periodicity* of the torsional term. For methanol, using $\alpha_0 = 0$, and $n = 3$, and including three such terms (one for each hydrogen bonded to carbon), will introduce the appropriate behaviour, with three minima where the oxygen-bound hydrogen is staggered with respect to the carbon-bound hydrogens.

For molecules with atoms that have a planar bonding environment, such as formaldehyde, another type of energy term is needed in order to describe the preferred planar structure at carbon: for this molecule, there will be three stretching terms,

---

A closer look at force constants

To evaluate the energy of the water molecule shown in the margin, you need to know the values of the force constants $k_{stretch}$ and k_{bend}, and of r_0 and θ_0, which describe the equilibrium structure. Typical values used for water are: $k_{stretch} = 1880\ \text{kJ mol}^{-1}\ \text{Å}^{-2}$, $k_{bend} = 0.070\ \text{kJ mol}^{-1}\ \text{degree}^{-2}$, $r_0 = 0.96\ \text{Å}$, and $\theta_0 = 104.5°$. Using $r_A = 1.24\ \text{Å}$ thereby yields a stretching energy of $1880 \times (1.24–0.96)^2 = 147.4\ \text{kJ mol}^{-1}$, r_B yields an energy of $1880 \times (0.91–0.96)^2 = 4.7\ \text{kJ mol}^{-1}$, and θ yields a bending energy of $0.07 \times (72–104.5)^2 = 73.9$ kJ mol^{-1}, yielding an overall potential energy $V = 226\ \text{kJ mol}^{-1}$. For this simple example, the equilibrium structure returns $V = 0$; this is not always true for complicated systems.

three bending terms, and one *improper bending* term, which is smallest when the atom involved has a planar structure. Various expressions can be used for this term.

As well as the stretching, bending, torsion, and improper torsion terms, more sophisticated forcefields can include additional cross-terms describing e.g. the modification of a stretching force constant upon bending. Including such terms can lead to more accurate forcefields, but it is notable that satisfactory results can be obtained even without them, suggesting that equations (4.1) and (4.2) provide a reasonable description of the underlying interactions.

Now consider interaction between several molecules, e.g. between two water molecules. Additional energy terms are required to account for the attractive and repulsive interactions. *Repulsion* arises whenever the electron clouds of the two molecules start to overlap significantly, and is due to the Pauli principle. This effect very rapidly becomes highly repulsive at distances below the threshold where the overlap begins. For interaction between two polyatomic molecules, one can reproduce this effect by considering one contribution from each pair of atoms AB, where A belongs to one molecule and B to the other. The contributions can be of the form shown in equation (4.3):

$$V_{repulsion}(r_{AB}) = \frac{A_{AB}}{r_{AB}^n} \qquad (4.3)$$

where A_{AB} is a constant measuring the strength of the repulsive interaction between the two atoms, and n is a large positive integer, which for historical reasons associated with computational efficiency in the early days of computational chemistry is often chosen to be equal to 12. Other numerical expressions are sometimes used in place of equation (4.3), e.g. an exponential function of the type $A_{AB} \times \exp(-\gamma_{AB}r_{AB})$, known as a 'Buckingham' potential after the British physicist and computer scientist Richard Buckingham.

There is also an *attractive* interaction between unbonded atoms, due to London dispersion effects. This can be represented by a sum of terms of the form:

$$V_{attraction}(r_{AB}) = \frac{-B_{AB}}{r_{AB}^m} \qquad (4.4)$$

In equation (4.4), B_{AB} is a constant measuring the strength of the attractive interaction for the two atoms considered, and m is a positive constant—which here is usually equal to six as this reproduces the known distance-dependence of dispersion interactions. The two terms in equations (4.3) and (4.4) are often grouped together and re-written as equation (4.5):

$$V_{Lennard\text{-}Jones}(r_{AB}) = 4\varepsilon\left\{\left(\frac{\sigma_{AB}}{r_{AB}}\right)^{12} - \left(\frac{\sigma_{AB}}{r_{AB}}\right)^6\right\} \qquad (4.5)$$

This functional form—as illustrated in Figure 4.1—is called the Lennard-Jones potential after the British physicist who invented it. The constant ε describes the strength of the interaction between the two atoms considered, and σ describes the cut-off distance at which the interaction changes from being negative, for large r_{AB}, to positive, at small r_{AB}.

For polar molecules, a forcefield typically also includes *electrostatic* energy terms, describing the interactions arising between ions and polar molecules due to Coulomb's law.

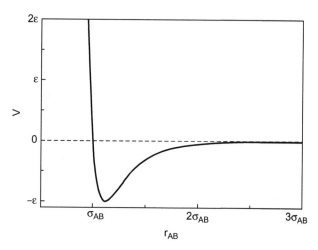

Figure 4.1 The Lennard-Jones potential.

The simplest way to achieve this is to assign *partial charges q* to each atom, and to include a sum of terms arising from Coulomb's Law to the potential energy:

$$V_{Coulomb}(r_{AB}) = \frac{1}{4\pi\varepsilon} \times \frac{q_A q_B e^2}{r_{AB}} \tag{4.6}$$

The partial charges q_A and q_B are fixed parameters, expressed in atomic units, e is the magnitude of the charge of an electron, and ε is the dielectric permittivity of the medium, typically chosen to be ε_0, the permittivity of vacuum. Such point charges allow forcefields to reproduce the structure and energetics of hydrogen bonds reasonably well (the positive hydrogen of the hydrogen-bond donor is attracted to a negative charge on the acceptor), although in some forcefields dedicated terms to describe hydrogen bonds are included.

Calculating the Coulomb potential energy: an example

Consider a sodium ion (with a charge of $+e$, i.e. 1.602×10^{-19} C) and a chloride ion (with a charge of $-e$), placed 5 Å apart in vacuum. The Coulomb potential energy arising as a result of their interaction can be calculated using equation (4.6) and the value for the dielectric permittivity of vacuum, $\varepsilon_0 = 8.854 \times 10^{-12}$ F m^{-1} i.e. 8.854×10^{-12} C^2 J^{-1} m^{-1}. The resulting $V = 1/(4 \times \pi \times 8.854 \times 10^{-12}) \times (1.602 \times 10^{-19})^2/(5 \times 10^{-10}) = 4.61 \times 10^{-19}$ J $= 278$ kJ mol^{-1}.

Coulombic interactions in molecular systems are more complex than allowed for in standard forcefields for two main reasons. First, the electron cloud in a given molecule or functional group adopts a complex three-dimensional distribution, whose electrostatic properties are not perfectly captured by a small number of atom-centred point charges. Greater accuracy is achieved by including dipoles, quadrupoles, and higher-order multipoles in general, centred either at the atomic positions or at special positions such as the centre of mass of the molecule, or the location of lone pairs. Also, the use of *fixed* partial charges (and multipoles) is an approximation: these charges should in fact change as molecules become more or less polarized due to their environment. For example, the electron cloud in a water molecule will distort in the presence of the electric field created by another water molecule, or by an ion or other solute. To describe this, one needs to allow q (as well as any dipoles, quadrupoles) to vary. This is the object of *polarizable* forcefields, which go beyond the scope of this work.

The Lennard-Jones and Coulomb terms (together referred to as the *non-bonded atom terms*) are usually omitted for pairs of atoms that are directly bonded to one another, or that are both bonded to the same atom. They are also either completely or partly omitted for atoms that are separated by three chemical bonds. For pairs of atoms that are in the same molecule, but further apart (e.g. the atoms in separate side-chains of a protein), they are however included, as these atoms interact in the same way as atoms in separate molecules.

Overall, the combined use of the simple expressions above for bond stretching, bond bending, dihedral angle torsion, improper torsion, and non-bonded interactions can lead to a surprisingly accurate description of the potential energies of molecules and collections of molecules. They can be used to predict the preferred structures of complex molecules based on parameters chosen based on properties of much simpler molecules, and also to predict relative energies of different conformers, and properties such as infra-red spectra. The way in which this is done will be described in Chapter 5. More sophisticated mathematical forms for each of the terms, and introduction of additional terms to describe e.g. the way in which stretching a given bond changes the energy cost of bending it, leads to improved accuracy but the simple form discussed here is often used.

4.3 Parameter sets

The forcefield energy expressions given above require many parameters, the force constants $k_{stretch}$ and k_{bend}, the equilibrium bond lengths r_0 and angles θ_0, the torsional constants $A_{dihedral}$ and associated n and α_0, the terms for improper torsion (not shown above in detail), and the non-bonded parameters σ, ε, and q. For a system with N atoms in total, there will typically be some multiple of N stretching, bending, torsion, and improper torsion terms, and some multiple of N^2 non-bonded terms. In principle, each of these terms could be assigned separate parameters to reflect the different chemical environment of each atom in the system. This would, however, be impractical in most cases. In practice, various procedures are used to minimize the number of parameters needed, and to derive appropriate values for these parameters from experiment or from quantum chemical calculations.

The main simplification, and one that is almost always used in forcefield development, is to introduce so-called *atom types*: atoms that share the same expected bonding and interaction properties. For example, in a protein, every carbon atom in the carbonyl group of every peptide bond could be assigned to the same atom type, and likewise for every peptide bond nitrogen. Then the same stretching force constant $k_{stretch}$ and equilibrium bond length r_0 can be used for every single C–N bond within a peptide bond—for example those highlighted here in the simple tripeptide Ala–Ser–Phe. The partial charge q_A on all atoms of the same type will be the same also. Use of atom types sacrifices some accuracy and specificity, since in principle each bond will have slightly different properties, each atom slightly different electrostatic properties, and so on. However, it also makes the number of parameters that need to be assigned much smaller, which makes parameterization much more simple.

The degree of detail in which such atom typing is carried out can vary. For example, if one wishes to develop a forcefield for alkanes, then various strategies are possible. The most aggressive simplification will result from choosing just two atom types, sp^3 carbon and hydrogen. In a more complicated but more accurate approach, one could treat methyl, methylene, methine, and quaternary group carbons as belonging to different atom types. Likewise, one could introduce different atom types for the hydrogen atoms in the different environments, or for carbon atoms within small rings such as cyclopropanes. Also, the atom types may have different scopes for different types of interaction, so that for example two atoms are considered to have the same type for the torsional energy expression, but different types for bond stretching. The choices made here tend to be slightly different from one forcefield to another.

A second simplification is the use of *combination rules* for the non-bonded Lennard-Jones interactions. If there are n_{type} atom types, then in principle you need $n_{type}(n_{type}-1)/2$ ε and σ parameters to describe all the mutual Lennard-Jones interactions. For example, a peptide bond carbon atom needs one σ parameter to describe its interaction with a carbon atom of a methyl group on an alanine residue, another for interactions with the hydrogen atoms of that methyl group, yet another for interaction with the nitrogen atoms of remote peptide bonds, and so on. This can be quite a lot of parameters, which in turn requires a lot of work in order to fit all the values. Instead, in most forcefields only the Lennard-Jones parameters σ_{XX} and ε_{XX} for interaction between two atoms with the same atom type X are defined. Then, the Lennard-Jones parameters for interactions between atoms of different types are derived from them. One popular choice, shown in equation (4.7), is to take the arithmetic average of the σ values, and the geometric average of the ε values:

$$\begin{cases} \sigma_{AB} = \dfrac{1}{2}(\sigma_{AA} + \sigma_{BB}) \\ \varepsilon_{AB} = \sqrt{\varepsilon_{AA} \times \varepsilon_{BB}} \end{cases} \tag{4.7}$$

This makes sense, since σ relates to the 'size' of the two atoms interacting with one another—which should quite naturally be roughly additive-and ε relates to the net effect of polarizability of the two partners, and should therefore depend on the product of two atomic terms. Other combination rules can also be used.

Once all of the choices above have been made—which terms to use in the forcefield, which precise energy expression to use, how to carry out atom typing and assign non-bonded parameters—one can then choose the parameters so as to complete the specification of the forcefield. This is not an easy task! It is usually carried out by using, as inputs, various pieces of *experimental* and *quantum chemical* information. On the experimental side, X-ray crystallography can be used to specify equilibrium bond lengths and angles, infrared spectroscopy can be used to assign force constants, and bulk properties, such as density or viscosity, can be used to help to assign non-bonded terms. Quantum chemistry offers an alternative way to calculate relative potential energies for different arrangements of atoms in a system, and forcefield parameters can be varied so as to reproduce as well as possible these relative energies.

In practice, fitting forcefields is complicated because one needs to establish priorities within the properties of the system to be described by the forcefield. For example, a forcefield that yields a good description of the hydrogen bonding pattern in liquid water may lead to less good predictions for the density as a function of temperature.

Another challenge is generality versus accuracy. There are many different molecules for which a user may wish to have a forcefield—many more than the number of systems for which a forcefield developer can carry out tests as part of the fitting procedure. If one wishes to have a very general forcefield, it may be desirable to assign many parameters even though very limited experimental or quantum chemical data was available to constrain the values adopted. The parameters in such cases may well be chosen based on little more than an (educated!) guess. On the other hand, if one wishes to have a very accurate forcefield for a very specific system, one may develop only parameters for that specific system.

Many forcefields are available to describe a wide range of molecular systems. Some are extremely well tested forcefields for specific systems—e.g. liquid water. Others are quite general forcefields that can describe quite diverse systems—e.g. the general biological forcefields such as AMBER (Assisted Model Building with Energy Refinement, developed initially by the American theoretical chemist Peter Kollman and co-workers) or CHARMM (Chemistry at Harvard Macromolecular Mechanics, developed by the American and Austrian theoretical chemist Martin Karplus and co-workers) that can describe pretty much any protein or nucleic acid molecule. Finally, others are very general forcefields that can describe *any* chemical system, such as the Universal Force Field UFF, developed by the American theoretical chemist W. A. Goddard III and co-workers. All of these forcefields can be useful for understanding molecular behaviour, but the user should remember that the quality of predictions will depend strongly on the accuracy of the forcefield, and especially the care which has gone into choosing its parameters.

4.4 Periodic systems and cut-offs

Molecular mechanics is mostly used nowadays to model the properties of very large systems, e.g. macromolecules, solids, liquids, or solutions. In fact, modelling of large biomolecules in aqueous solution is probably one of the largest fields in computational science, based on the number of hours of computer time allocated to it. When treating large systems, there are a number of challenges associated with the efficiency of the calculation. For example, for a system with N atoms, leaving aside the small number of atom pairs for which non-bonded interactions are omitted, there are about $N^2/2$ non-bonded terms to calculate. For a system with 10,000 atoms, that is of the order of 50,000,000 Lennard-Jones and Coulomb terms to evaluate! Yet many of these pairwise interactions are very small.

Just how small are non-bonded interactions?

To see how small many of the non-bonded interactions in a molecular forcefield will be, one can do a rough calculation. Consider a typical system, a protein or protein fragment with a total of 10,000 atoms. Assume that the density of the system is 1 g cm^{-3} and that the average mass of the atoms in the system is 8 amu. The system as a whole will have a mass of 80,000 amu or 1.33×10^{-22} kg so it will have a volume of 1.33×10^{-25} m^3. If it is a sphere, its volume is $4\pi r^3/3$, so the radius would be 3.17×10^{-9} m, or 32 Å. Many of the atoms will be at least this far from each other. Consider two atoms, each with a partial charge of half an electron (0.8×10^{-19} C), and interacting through a Lennard-Jones potential with $\sigma = 3.2$ Å and $\varepsilon = 0.6$ kJ mol^{-1}. Their Lennard-Jones interaction energy will be of -2.4×10^{-6} kJ mol^{-1}, and their Coulombic interaction energy will be of 1.8×10^{-20} J or 11 kJ mol^{-1}.

One way to lessen the computer time needed to evaluate the molecular mechanics energy of very large systems is to use cut-offs in energy expressions. For an interatomic distance greater than a threshold, non-bonded terms can be simply omitted. Because this can lead to discontinuities in the energy surface, modified cut-off schemes can be used. One possibility is to shift the potential energy term upwards by its value at the cut-off distance, and to again omit this term for distances larger than the cut-off, as shown in equation (4.8):

$$V_{cutoff}(r) = \begin{cases} V_{standard}(r) - V_{standard}(r_{thresh}) & \text{if } r \leq r_{thresh} \\ 0 & \text{if } r \geq r_{thresh} \end{cases} \tag{4.8}$$

This yields a continuous energy expression, but the derivative of the energy with respect to atomic coordinates will still be discontinuous. Yet another option is to 'smooth off' the transition by multiplying the non-bonded term by a switching function $f_{switch}(r)$:

$$f_{switch}(r) = \begin{cases} 1 & \text{if } r \leq r_{thresh1} \\ g(r) & \text{if } r_{thresh1} \leq r \leq r_{thresh2} \\ 0 & \text{if } r \geq r_{thresh2} \end{cases} \tag{4.9}$$

In equation (4.9), the function $g(r)$ has a value of 1 at the lower threshold distance, and a derivative of zero, then decreases continuously to zero at the higher threshold distance, where its derivative is also zero. In this way, the interaction is continuous across the whole range of distances, and so is the derivative.

For the Lennard-Jones term, the use of a cut-off may appear reasonable, since for large enough thresholds, each individual interaction is very small, as shown in the worked example above. It should, however, be noted that as one moves farther away from a given central atom, there are ever more atoms that are roughly the same distance away. If the mean atom number density in the system is ρ, then there are $(4\pi r^2 \rho \, dr)$ atoms at a distance between r and $r + dr$ from a given central atom. For the same example given above, with 10,000 atoms in a volume of 1.33×10^{-25} m^3, this means that the small shell situated between 31.9 and 32 Å from the centre of the system will contain roughly 100 atoms. So the neglect of the Lennard-Jones terms can be substantial in total if the cut-off used is too short.

For ionic interactions, individual interactions will be much larger for a given cut–off distance–as shown above. However, unlike the Lennard-Jones terms, which are all *attractive* at large r, coulombic interactions may be repulsive or attractive, and the $(4\pi r^2 \rho \, dr)$ atoms within a shell of thickness dr at a distance r from the central atom will have a mixture of positive and negative charges, so that, on balance, this shell of atoms will be roughly neutral. This means that attraction and repulsion to the charge on the central atom can be expected to cancel out, at least roughly so. Because of this, it still makes sense to use cut-offs even for coulombic interactions.

Even using cut-offs, it is difficult to build models that are large enough to represent bulk behaviour of liquids or solids in a meaningful way. For this to be the case, *most* of the atoms need to be *far away* from the surface of the system being modelled. What exactly is meant by 'most' and by 'far away' may well depend on what one is trying to model, and how accurate one wants the answer to be. Still, pretty much any reasonable choice for both terms will mean that the modelled system needs to contain many

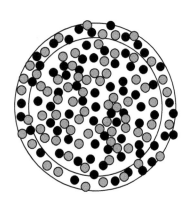

Fraction of bulk-like atoms in a sphere of water molecules

Imagine a spherical droplet of water with radius r and a standard density of 1 g cm^{-3}. Using a molar mass of 18 g mol^{-1}, and Avogadro's constant of 6.022×10^{23} mol^{-1}, that corresponds to a density of 0.033 molecules per Å^3, or a mean volume per water molecule of 29.9 Å^3. Assume that water molecules are roughly spherical; the diameter of a sphere with volume 29.9 Å^3 is 3.85 Å. Now, let us define 'far away' from the surface as being further away from the surface of the droplet than the diameter of a water molecule, and let us take 'most' to mean more than 3/4 of the molecules. What is then the minimum radius r_{min} for a water droplet that would have 'most' molecules 'far away' from the surface? With these definitions, we must have that $(r_{min} - 3.85 \text{ Å})^3/(r_{min})^3 >$ 0.75, yielding $r_{min} = 42.1$ Å. That corresponds to a system containing about 2500 water molecules (7500 atoms), with just under 1900 of the molecules 'far away' from the surface. Other definitions for 'most' and 'far away' will likewise lead to the conclusion that many atoms are needed.

thousands of atoms before it starts to display bulk-like behaviour—and that may be too many for the desired calculations to be feasible.

Hence another technique is needed to allow very extended systems to be modelled without needing punitively large numbers of atoms to be considered: *periodic boundary conditions*. These were already discussed in a rather different context, in section 3.10 on quantum chemistry modelling for periodic systems. With periodic boundary conditions, one assumes that one is modelling an infinitely sized system, formed by an infinite number of translationally symmetric unit cells. Another way of viewing this is that one has one 'main' simulation system, surrounded by an infinite number of periodic images of itself. In the simplest cubic arrangement, for any atom with coordinates (x, y, z) within the central system, there are also equivalent atoms with coordinates $(x + n_x L, y + n_y L, z + n_z L)$, where L defines the size of the cubic images, and n_x, n_y, and n_z can be any positive or negative integer. This is shown schematically in two dimensions in Figure 4.2 (only a few images are shown).

This may appear to introduce an impossible level of complication to the system, since there are now an infinite number of atoms to model! However, combined with the use of cut-offs, it is possible to calculate the energy of such a system while

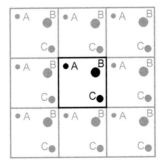

Figure 4.2 Schematic representation of how the multiple images of a system relate to each other when using periodic boundary conditions. The central image is shown in black, with a few of the surrounding images shown in grey. A, B, and C are three generic atoms in the system.

including only roughly the same number of energy terms as are needed to describe a single cubic image (or replica). This is done by considering the 'main' replica of each atom (shown here in black) in turn. Only the interactions involving the *nearest* replica of each of the other atoms are included in the energy expression. For example, consider that there is a bond, with a stretching term, between atoms A and B. Consider the main replica of A, in the central system image. The B replica that will be used to compute the stretching term will be the one in the system image situated to the left of the central image, as that is the B atom that is closest to the starting A atom. If there is a bending term of the type C–A–B, then this will be computed using the C atom in the top left image, the A atom in the central image, and the B atom in the left image, because this choice corresponds to the shortest C–A and A–B distances. One could equally choose the C atom in the central image, the A atom in the lower right-hand side image, and the B atom in the bottom image—but this would yield the same energy contribution, because each image is an exact replica of the other, and the images are disposed in a periodic array, so the corresponding angle would be exactly the same.

For non-bonded interactions, one calculates them, likewise, by considering the nearest image atom. In the present example, assuming that there are no bonds, one needs to consider terms arising from pairs of atoms A and B, B and C, and A and C. As in the examples given above, these terms will be calculated using the appropriate atom, taken either from the same image of the system, or from a neighbouring one. It turns out that it is also necessary for non-bonded interactions to use a cut-off distance, which must be smaller than half of the box length L. For the Lennard-Jones terms, this is straightforward to do, as described above. For coulombic interactions, which act at longer range, a complete neglect beyond a certain distance may be considered inappropriate. A more sophisticated approach called *Ewald summation* exists for these terms, to take into account longer-range interactions also, but this goes beyond the scope of this book.

4.5 Practical aspects of molecular mechanics methods

As for all the methods described in this book, most computational projects based on molecular mechanics will require the use (or development) of quite a complicated computer program. There are quite a few of these for molecular mechanics. Almost all of these programs will be programmed to explore the molecular mechanics potential energy surface in many of the ways described in Chapters 5 and 6, and some examples will be given there. Some of these programs will also be able to carry out quantum chemical calculations using the methods described in Chapters 2 and 3, and also to carry out multiscale calculations as discussed in Chapter 8.

One of the most difficult tasks in setting up a molecular mechanics problem is to choose and specify the aim of the calculations, and the model on which calculations are to be performed. When setting the aim of a molecular mechanics project you need to bear in mind what MM can and cannot do. It can be used to calculate the energy for a set of structures. If they all can be described using the same forcefield, then the *relative* energies can be compared and conclusions drawn about stability. Also, as discussed in Chapter 5, you can locate optimum structures, and perhaps reaction paths. However, it is not possible to describe changes in chemical bonding: this is assigned once and for all at the start of an MM study. So the 'reactions' one can describe are only changes in conformation, or changes in non-covalent 'bonding' of a type that can be described by non-bonded MM terms. If it is important to study changes in bonding, then MM is not the right method. MM can also be used in conjunction with the simulation methods of Chapter 6 to describe behaviour of a system at a given temperature.

MM calculations can be useful for small molecular systems, usually as part of a more complex computational study. More typically, though, they are used to study very large systems, which is made possible thanks to the computational efficiency of MM. Broadly speaking, a system with N atoms will contain of the order of N bond stretching terms, angle bending terms, and dihedral terms. It will contain of the order of N^2 non-bonded terms (charge–charge interactions and van der Waals). If one uses cut-offs as discussed in section 4.4, then many of these non-bonded terms do not need to be considered, and for large N, the computational effort needed to evaluate those that remain is again some multiple of N. Overall, the computing time for MM methods scales roughly linearly with the number of atoms. By using periodic boundary conditions, one can model condensed phase systems in a realistic way.

One of the biggest challenges with such large systems is to set up and monitor the simulations. Each atom needs to be assigned a set of bonds, and parameters for charge, van der Waals, etc. interactions. This is made easier through the use of atom types as discussed in section 4.3 but it remains challenging, and set-up often requires some degree of automation. For example, for proteins or other biomolecular systems, great care is needed to assign each acidic or basic residue with the correct charge state, and the sheer overall number of atoms means that mistakes can sometimes be made and not immediately detected.

4.6 Further reading

The general books listed under this heading in Chapter 1 will provide further information on molecular mechanics forcefields. Here are three classic papers from different decades reporting progress in forcefield design.

- Energy Functions for Peptides and Proteins. I. Derivation of a Consistent Force Field Including the Hydrogen Bond from Amide Crystals. A. T. Hagler, E. Huler and S. Lifson, *Journal of the American Chemical Society*, 1974, **96**, 5319–5327.

- Development and testing of the OPLS all-atom force field on conformational energetics and properties of organic liquids. W. L. Jorgensen, D. S. Maxwell and J. Tirado-Rives, *Journal of the American Chemical Society*, 1996, **118**, 11225–11236.

- Automation of the CHARMM General Force Field (CGenFF) I: Bond Perception and Atom Typing. K. Vanommeslaghe and A. D. MacKerell, Jr., *Journal of Chemical Information and Modeling*, 2012, **52**, 3144–3154.

4.7 Exercises

4.1 Using a molecular mechanics program, carry out a calculation for a simple system such as a single water molecule (most programs will provide a small number of 'test jobs', one of these would be ideal for the present purpose). Identify the energy terms included in the forcefield, including the parameters.

4.2   For a more complicated system such as a small peptide, identify the energy terms and the atom types assigned to each atom and bond.

4.3   Again using a sample input file, identify how periodic boundary conditions are specified in the input for a calculation on a bulk system. Also identify the keywords that request the use of cut-offs, and test their effects.

4.8 **Summary**

- The potential energy surface of molecules, and of collections of molecules, can be quite well described using a 'mechanical' description with atoms as 'balls' connected by 'springs'.

- Forcefields typically contain energy terms for bond stretching, angle bending, dihedral torsions, and non-bonded interactions, including charge–charge and van der Waals forces.

- Parameter sets suitable for describing many classes of molecules and biomolecules have been developed and tested.

- Efficient calculations on realistic large models of condensed-phase systems, such as macromolecules in solvent, requires the use of cut-offs to restrict the number of non-bonded terms.

- Periodic boundary conditions are frequently used so as to be able to represent bulk-like behaviour of condensed-phase systems.

5 Geometry Optimization

5.1 Introduction

In Chapters 2, 3, and 4, we considered the most important computational methods for calculating the energy of a chemical system for a given set of positions of the nuclei of the atoms making up the system. Indeed, within the Born–Oppenheimer approximation, there is a unique ground-state energy for each arrangement of the nuclei. However, for different positions of the atomic nuclei, the energy is different: it is a function of the set of all the atomic coordinates R. This energy function or energy surface, called a *potential energy surface*, can be used to predict the chemical behaviour of the system. In this chapter, we will describe the features of potential energy surfaces, with a focus on energy minima and saddlepoints or transition states. We will also describe methods to explore potential energy surfaces in a systematic way through *geometry optimization*. In the partner Chapter 6, methods to *simulate* the motion of atoms on energy surfaces will be described.

5.2 Features of potential energy surfaces

The ground-state potential energy surface of a system can be referred to as the function $V(R)$, where R is a vector containing the coordinates of all the atomic nuclei in the system, and V is the ground state electronic energy for the system, as obtained from solving the electronic Schrödinger equation, or by evaluating a molecular mechanics forcefield expression. In general, for a system with N atoms, $V(R)$ is a function with $3N$ variables (or degrees of freedom), since each of the atoms can move in the x, y, and z directions. However, some combinations of these motions lead to no change in V: those that preserve the internal structure of the system. In the absence of interactions with the environment, simply translating the considered molecule or collection of molecules through space does not change V, and nor does rotating it. Hence six degrees of freedom (three translations and three rotations) do not affect V, which in practice therefore has $3N - 6$ degrees of freedom. Diatomic molecules ($N = 2$) are a special case, since they can only rotate around two distinct axes, so V depends on $3N - 5 = 1$ coordinate, the interatomic distance.

Simple representations of potential energy surfaces are often used in chemistry. A typical example is shown in Figure 5.1 for a generic transformation involving reactants converting to products through an intermediate and two transition states.

In Figure 5.1, the low-energy 'wells' labelled '**Reac**', '**Int**' and '**Prod**' correspond to 'stable' species that persist for a certain period of time. If the well they reside in is deep, then the corresponding lifetime may be long enough that the species can be isolated

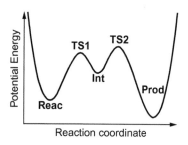

Figure 5.1 Schematic plot of V(R) for a reaction leading from **Reac** to **Prod** through **TS1**, **Int**, and **TS2**.

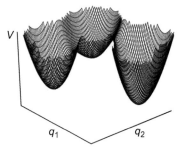

Figure 5.2 Perspective representation of a potential energy surface, showing dependence of V(R) on two degrees of freedom.

and characterized at least under appropriate conditions. The high-energy points TS1 and TS2 are transition states, critical points that must be passed through when **Reac** converts to **Int** or **Int** to **Prod**.

The plot shown in Figure 5.1, while very useful, is also highly simplified, since it shows only one of the $3N - 6$ coordinates. Plotting a potential energy surface with all dimensions is impossible (except for diatomic molecules). A slightly less simplified representation of the same system can be obtained by using perspective to show three dimensions, i.e. to plot the potential energy V as a function of two coordinates (shown in Figure 5.2 as the generic q_1 and q_2 coordinates).

Rather than using perspective, the same surface can also be represented (Figure 5.3) on the two dimensions of a page or screen using a contour plot (q_1 and q_2 are again the two coordinates).

Even the representations of Figures 5.2 and 5.3 are still hugely simplified, since only two degrees of freedom are shown. Nevertheless, they do allow some additional and important features of potential energy surfaces to be shown, compared to Figure 5.1. First, the low-energy regions labelled 'Reac', 'Int' and 'Prod' are now seen to lie at local minima *in all directions* on the potential energy surface, with the energy rising upon undergoing small distortions in either the q_1 or the q_2 direction. The TSs, which appear as local maxima in Figure 5.1, are indeed maxima in one direction—the reaction coordinate—but minima in the other direction. In general, for an N-dimensional

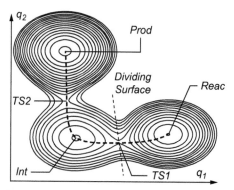

Figure 5.3 Contour plot representation of a potential energy surface, showing dependence of V(R) on two degrees of freedom. The positions of the minima and TSs (or saddlepoints) is indicated, as well as the dividing surface between the two neighbouring wells to TS1.

potential energy surface, transition states correspond to first-order saddlepoints, which are maxima in one direction, but minima in the $N-1$ other ones. The reason why such saddlepoints play such an important role in chemistry is because they represent the lowest energy points within the $(N-1)$-dimensional dividing surface between two low-energy regions of the potential energy surface. Incidentally, reactions do not require that the system pass exactly through the saddlepoint, although they do need to pass the dividing surface at some point near the saddlepoint. Positions on the potential energy surface that are maxima in more than one direction exist, but they do not play a very important role in chemistry since there are always lower-energy routes for moving from one low-energy region to another that skirt around such points.

Another benefit of multi-dimensional plots such as in Figure 5.2 or 5.3, compared to the one-dimensional curve in Figure 5.1, is that they illustrate an important property of the lowest energy path (or reaction path, shown as a dashed line in Figure 5.3) leading from **Reac** to **Prod**. This path does not typically involve only changes in one coordinate throughout the reaction: the first step, from **Reac** to **Int**, involves mostly change in q_1, whereas the second step involves mostly q_2. Also, for each of the steps, the minimum energy path is somewhat curved.

The figures discussed in the paragraph above provide a schematic representation of potential energy surfaces for systems where the reactant, intermediate, and product all correspond to molecular species containing the same number of atoms. In other words, the reactions illustrated convert *one* molecule (a minimum on the potential energy surface) into one of its isomers (a different minimum). Such steps are also referred to as *unimolecular*. As examples of unimolecular reactions, consider for example the Cope rearrangement of 1,5-dienes or the ring opening of a cyclobutene shown opposite.

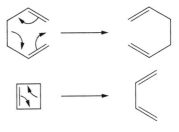

Many important reactions are instead *bimolecular*, in that they convert two molecules (or molecular fragments) into one (or two) molecules. Equations (5.1) to (5.5) provide some examples of such reactions:

$$H + F \rightarrow HF \tag{5.1}$$

$$CH_3 + OH \rightarrow CH_3OH \tag{5.2}$$

$$H_2O + C_6H_5OH \rightarrow C_6H_5OH \cdot H_2O \tag{5.3}$$

$$CH_4 + OH \rightarrow H_2O + CH_3 \tag{5.4}$$

$$CH_2 = CH_2 + CH_2 = CH - CH = CH_2 \rightarrow \text{cyclohexene} \tag{5.5}$$

The simplest example of this type is the combination of two atoms A and B to form a diatomic molecule AB, as in equation (5.1). As mentioned above, the potential energy surface for this simple process really does only depend on one coordinate, the interatomic distance, so a one-dimensional plot (Figure 5.4) is adequate to show the whole surface. Note that the 'reactant' state A + B does not correspond to a minimum here or even to a particular structure.

As shown in Figure 5.4, the schematic 'reaction' of atoms A and B to form molecule AB does not involve a potential energy barrier. Almost all atom–atom potential energy surfaces share this property. For bimolecular reactions between molecules or molecular fragments, this is however not always the case. Where 'A' and 'B' are molecular fragments with unfilled valencies (radicals) on one of their atoms (e.g. atom X in A and Y in B), a cut through the potential energy surface along the X–Y distance will typically also look like Figure 5.4, provided that the two radicals approach each other with a

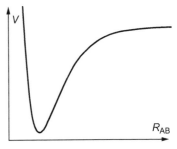

Figure 5.4 Potential energy curve for a diatomic molecule AB, showing V as a function of the internuclear distance R_{AB}.

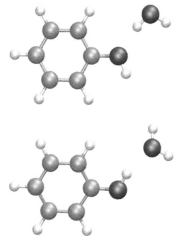

Figure 5.5 Structure of two isomers of the complex between phenol and water.

relative orientation similar to the one that they will adopt in the final molecule AB. An example would be the potential energy surface for equation (5.2), showing the energy as a function of distance between the C atom in CH_3 and the O atom in OH as they combine to form methanol.

When two stable molecules interact, they can usually also form a complex which is more stable than the two separated species, due to attractive interactions such as hydrogen bonding, dipole–dipole interactions, or other van der Waals interactions. Reaction (5.3) between phenol and water is an example of such a process, leading to formation of a hydrogen bond. As for the case of two radicals interacting, such non-bonded interactions are also anisotropic: the two species need to be oriented in a way that favours the non-bonded interaction in order for the interaction to be uniformly attractive en route to the bound complex, as in Figure 5.4. Optimum structures for the product of reaction (5.3), the phenol–water complex, are illustrated in Figure 5.5. Depending on the relative orientation, there can be several different optimum geometries, two examples of which are shown. These two structures will be connected by a transition state. Such molecular complexes are almost always much less strongly bound than covalent molecules such as methanol.

In contrast with the simple atom- or radical-recombination processes discussed above, many bimolecular reactions involve an energy barrier, due to the fact that bond-making is accompanied by bond-breaking, as well as by non-bonding repulsion between the two partners. This is the case for the reactions in equations (5.4) and (5.5), and a schematic contour plot of the potential energy surface for a reaction such as equation (5.4) is shown in Figure 5.6. In such cases, there will again be a saddlepoint on the potential energy surface en route from reactants to products, as in Figure 5.2 or 5.3, but there will no longer be a minimum corresponding to the reactants, with the energy instead reaching an asymptote as the distance between the two reactants (q_1 in Figure 5.6) increases to infinity. In many reactions—such as equation (5.4), the atmospherically important activation of methane by the hydroxyl radical—there is also no product minimum, and the energy instead heads to an asymptote as the distance between the two product fragments (q_2 in Figure 5.6) increases. One can however consider separate, lower-dimension, potential energy surfaces for each of the reactants and products (in the example above, the four species CH_4, OH, H_2O, and CH_3), and these will display minima corresponding to the stable structures of these separated species.

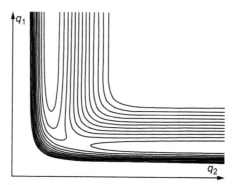

Figure 5.6 Contour plot representation of a potential energy surface for a bimolecular reaction.

The description on the previous page identifies the important features that occur on potential energy surfaces for chemical systems. Many *combinations* of the features mentioned above can also occur. For example, Figure 5.4 is a good qualitative representation of the potential energy for the different phenol–water hydrogen-bonded complexes of Figure 5.5 as the distance between the hydrogen-bonded atoms decreases. However, the two isomeric hydrogen-bonded complexes could also undergo a rearrangement leading from one isomer to another (and there will also be still more isomers, not shown in Figure 5.5). This part of the potential energy surface describing the isomerization will resemble Figure 5.2 or 5.3. As another example, the Diels–Alder reaction between butadiene and ethene, equation (5.5), resembles Figure 5.6 on the reactant side, for large distances q_1 between the two reacting molecules. On the product side, though, there is just one molecule, cyclohexene, residing in a minimum of the potential energy surface, somewhat like in Figure 5.2 or 5.3.

Excited-state potential energy surfaces

Most of the focus in this chapter will be on the *ground-state* potential energy surface, formed from the energy of the lowest-energy solution to the Schrödinger equation at a given set of atomic coordinates. For many systems, the energy separation between the electronic ground state and the excited states is quite large, and the chemical behaviour of the system can be rationalized purely in terms of motions on the ground-state potential energy surface. However, in some cases, the energy gap between ground and excited states is smaller, and indeed, for some arrangements of atoms referred to as *conical intersections*, it can drop to zero, so that two or more electronic states have the same energy. In such cases, important in photochemistry among others, one needs to take into account multiple potential energy surfaces and the regions where they intersect. This aspect goes beyond the scope of this primer.

Characterizing potential energy surfaces

Now that we have a good understanding of the nature and shape of potential energy surfaces *in general*, the question arises: how can we identify the features mentioned above–such as wells corresponding to stable species, saddlepoints corresponding to TSs, reaction paths, or asymptotes for separate species–for a specific system of interest?

The most obvious answer to this question might be simply to compute the value of the potential energy surface for a large number of different positions of the atoms making up the system, and then plot the potential energy function graphically. This is certainly possible for some small molecules. Some examples were shown in Chapter 3, such as Figures 3.4, 3.6, and 3.7. Another example is shown in Figure 5.7 for the hydrogen fluoride molecule, where the energy has been calculated using the CASSCF method at 13 structures with the H–F distance varying between 0.5 and 4.0 Å. As was discussed in Chapter 3, CASSCF is able to produce reasonable results even at large bond length, where the bond is broken, whereas HF or other correlated methods such as CCSD(T) would only be meaningful near the minimum. The set of values of the potential energy obtained is plotted using the diamond symbols shown; the (straight) lines connecting these points have been added merely so as to guide the eye. It is clear to recognize that

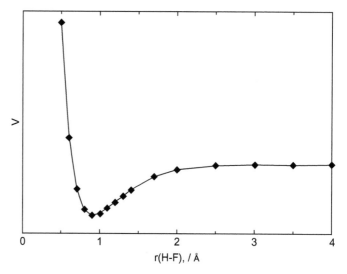

Figure 5.7 Systematic exploration of the potential energy surface for hydrogen fluoride, showing CASSCF quantum chemical energies calculated at 13 different values of the H–F distance r(H–F).

the potential energy surface resembles that of Figure 5.4, showing that HF is bonded, and has a stable structure with r(H–F) roughly equal to 0.9 Å. The energy then rises smoothly until an asymptote corresponding to separate F and H atoms.

A similar approach can be used for a molecule with three atoms, such as water. However, this molecule now has a potential energy surface with three degrees of freedom: the two O–H distances $R(O–H_1)$ and $R(O–H_2)$, and the H–O–H angle. At most we can display a *cut* through the whole potential energy surface, for example we can consider a set of different values of $R(O–H_1)$ and $R(O–H_2)$, while holding the angle fixed. Figure 5.8 uses perspective to illustrate the calculated potential energy (again obtained with CASSCF) for different O–H$_1$ and O–H$_2$ distances each ranging from 0.7

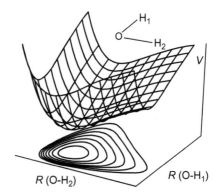

Figure 5.8 Systematic potential energy surface exploration for a water molecule, showing CASSCF quantum chemical energies for 121 different combinations of the two O–H distances (the angle is the same for all points). The potential energy surface is also shown as a contour plot below the surface.

to 1.7 Å in steps of 0.1 Å. The nodes in the wire-mesh used to display this surface are the points where the energy was computed—again they are connected by lines to guide the eye. For this simple system, it can be seen that the minimum, corresponding to the stable structure of water, is near $r(O–H_1) = r(O–H_2) = 0.95$ Å, while increasing one or other bond-distance leads to an increase in energy, that is starting to approach a plateau corresponding to $OH + H$ for the largest value plotted here. Increasing *both* O–H distances leads to a greater rise in energy, since it leads to formation of $O + H + H$.

The method applied above is not one that can be routinely used, for a number of reasons. First, this approach requires a large number of evaluations of the energy, which may require too much in the way of computational time. For H–F, thirteen energies were obtained, while $11 \times 11 = 121$ were used for water (though symmetry meant that some of these did not require separate calculation; only 66 unique energies were needed). In general, for a system with n degrees of freedom, a rough characterization of the potential energy surface would require considering ten or more values of each coordinate, and all combinations of values—i.e. of the order of 10^n structures. Even if the method used to calculate the energy is not particularly computationally demanding, the exponential growth in the number of points needed makes full mapping of the surface in this way impossible. Second, even if it were possible to calculate the energy at the large number of points needed, one would still face a problem with what to do with the obtained values: graphical plotting is only possible for one or two degrees of freedom.

Accordingly, other techniques are used to explore potential energy surfaces. These break down into two types: *static* methods, in which one again focuses on calculating the potential energy surface, and locating important parts of it; and *dynamic* methods, in which one seeks to mimic the motion of atoms in molecules, i.e. motion on the potential energy surface. Static methods can also be described as methods for *optimizing* structures, and will be the topic of the remainder of this chapter, with dynamic methods being covered in Chapter 6.

5.3 Geometry optimization methods

The most straightforward way to explore potential energy surfaces is to characterize them in terms of their stationary points, where the *gradient* of the potential energy with respect to the atomic coordinates vanishes. These are the minima and saddle-points mentioned above. In order to locate these points, numerical methods for 'geometry optimization' are used. As well as minima and saddlepoints, such methods can also be used to explore the reaction paths that connect them.

Optimization of a local minimum on the potential energy surface is the simplest problem of this type. A starting structure, i.e. a starting set of values of the atomic coordinates R, is generated for the system being considered, and the aim is to find the nearby structure R_{eq} for which $V(R)$ is a local minimum. Many algorithms are known for locating minima of functions. One option is to make a random change in the coordinates, ΔR, compute $V(R+\Delta R)$, and check to see if it is lower than the previous value of V. If yes, the change in structure is 'accepted', $R = R + \Delta R$, and one then repeats the process. If instead the energy is higher than before, then the change is rejected, and one simply repeats the process without updating the structure. Provided that the maximum magnitude of ΔR is chosen to become smaller and smaller as one nears the minimum, this algorithm will eventually lead close to R_{eq}. However, it is rather inefficient, and in practice this approach is very seldom used.

The method just described only requires that one be able to calculate the potential energy at a given point. More efficient algorithms generally require that one also be able to calculate the gradient $\partial V/\partial R$. For most quantum mechanical and molecular mechanical techniques, this can indeed be done, with a relatively small additional computational effort above that needed to compute the energy itself. A simple procedure for geometry optimization based on availability of the gradient is the so-called *steepest descent* method. An initial structure R_0 is chosen, and the potential energy $V(R_0)$ and its gradient $\partial V(R_0)/\partial R$ are computed at that point. The lowest-energy structure along the linear direction defined by the point R_0 and the direction of the gradient is then located. This is called a 'line search'. The potential energy and gradient at this new structure, R_1, is then obtained, and a new line search is performed along the direction defined by the new gradient. After several steps, this method, which involves repeatedly going as far 'downhill' as possible along the direction of the gradient—the steepest direction—will get close to the nearest minimum on the surface. The minimization will be considered to be converged when the size of the step taken and/or the magnitude of the gradient become smaller than some threshold. (Note that such 'convergence criteria' or cut-off values will also be needed for each of the line searches: the cut-off values for the line search can be less demanding than for the overall minimization.)

The typical behaviour of the steepest descent method is shown in Figure 5.9; the positions of the initial structure R_0 and ten subsequent structures R_1 to R_{10} are shown as dots, the exact position of the overall minimum (which here was known—this is not typically the case when carrying out geometry optimization!) is shown as a square. The minimization was considered completed because the gradient had dropped

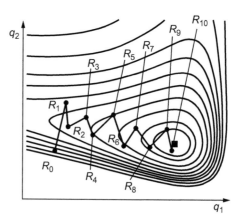

Figure 5.9 Schematic illustration of the steepest descent optimization method. The two-dimensional potential energy surface depending on q_1 and q_2 is shown as a contour plot. The starting point of the optimization and ten subsequent points visited during the optimization are shown as black dots, with the direction of each of the line searches shown as heavy lines. The position of the minimum is shown by a square.

below a certain target threshold. Note that the last dot is not exactly at the minimum: this is inevitable in numerical procedures such as minimization. Careful choice of convergence thresholds will however ensure that the structure obtained is sufficiently close to the minimum for practical purposes. Note also that for this simple system with only two coordinates, the steepest descent method requires ten steps (as well as the intermediate steps during the line searches, of which there may be 2–3 in each case). This seems rather a lot for such a simple problem, and is due to the fact that the steepest descent method does not 'learn' anything about the shape of the surface from one line search to the next. It just goes downhill along the steepest direction found at the latest point R_i.

More accurate methods for geometry optimization use not only the gradient—the first derivative of the energy with respect to the coordinates—but also the Hessian—the second derivative of the energy, shown in equation (5.6). This Hessian is a matrix, since for n different coordinates $q_1 \dots q_n$, there will be $n \times n$ matrix elements of the form:

$$H_{ij} = \frac{\partial^2 V}{\partial q_i\, \partial q_j} \tag{5.6}$$

It is possible to generate a much better estimate of the position of the minimum given an initial position R_i and the gradient and Hessian at that position. This estimate is obtained from Newton's method (equation 5.7, where $\mathbf{H}^{-1}$ is the inverse of the Hessian matrix):

$$R_{i+1} = R_i + \Delta R = R_i - \frac{\partial V(R_i)}{\partial R} \times \mathbf{H}^{-1} \tag{5.7}$$

Applying Newton's method in practice is challenging, because evaluating the full Hessian matrix is computationally demanding, and evaluating its matrix inverse can lead to numerical problems. Hence Newton's method is rarely applied as such in computational chemistry. However, many other so-called 'second-order' methods are used. These methods are referred to as 'second-order' because they in some way take account of the second derivatives of the potential energy, usually in some approximate way that avoids the need to calculate the exact Hessian matrix. There are too many such methods to describe them all in detail here! One example is the 'conjugate gradient' method, which is similar to the steepest descent approach in that it also involves line searches. However, it does not necessarily follow the steepest downhill direction for these searches, with the direction being instead modified so as to take account of the outcome of the previous steps. Another example of such a second-order method was first suggested by researchers Broyden, Fletcher, Goldfard, and Shanno and is therefore called the BFGS method. It uses equation (5.7) to update the structure at each step, but instead of evaluating the exact $\mathbf{H}^{-1}$ matrix, an approximate matrix computed from the value of the gradient at the previous steps is used. Each application of equation (5.7) yields a new structure with the outcome shown in Figure 5.10 for the same simple two-dimensional problem as shown in Figure 5.9. After just four steps, it reaches much closer to the minimum than the steepest descent method did in ten steps (the black square denoting the minimum has been removed as it would hide R_4).

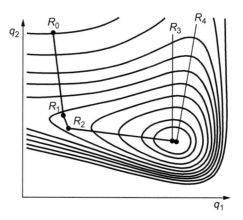

Figure 5.10 Schematic illustration of the BFGS optimization method, for the same model problem as in Figure 5.9.

5.4 Geometry optimization with quantum chemical methods

By and large, geometry optimization proceeds in the same way and uses the same algorithms whatever the underlying method used to calculate the potential energy surface: quantum chemistry, molecular mechanics, or other techniques. There are, however, some differences in detail. For geometry optimization of small systems of up to a few hundred atoms, such as is commonly needed when performing quantum mechanical studies, BFGS or closely-related methods are most often the best approach. To minimize the computational time needed to carry out the optimization, one needs to minimize the number of times that the quantum mechanical energy and gradient are evaluated. The computational challenge associated with evaluating the changes to the approximate Hessian, or with working out the Newton step of equation (5.7), is completely negligible in comparison to that of the quantum mechanical part of the calculation. In contrast, evaluating the *exact* Hessian matrix **H** quantum chemically for any given set of coordinates is quite demanding in terms of computer time. In some cases, though, the additional accuracy compared to the use of an approximate **H** as in the BFGS method can be significant. For example, the number of structures R_i that must be considered prior to reaching an acceptably converged optimum structure can be much reduced. Hence the Hessian matrix is sometimes calculated quantum chemically a few times during the optimization.

Another factor that can improve convergence of the optimization procedure is careful choice of the set of coordinates q that are used to describe the structure of the system. One obvious choice is simply the **Cartesian coordinates** of the atoms. These have the merit of leading to simple algorithms, because the gradient of the energy is most readily obtained with respect to these coordinates, whether one is using a molecular mechanical or a quantum mechanical potential energy method. However, the use of Cartesian coordinates does not always lead to a smoothly converging optimization. This is because a small change of a given Cartesian coordinate can lead either to a large change in energy, if it is associated with a change in a chemical bond length, or to a small change, if it instead corresponds merely to a change in dihedral angle. Also, each Cartesian coordinate typically maps onto changes in many different such bond lengths or angles, which leads to complicated interdependence of the different coordinates concerning their effect on the potential energy, and thereby to slower convergence.

Internal coordinates (bond lengths, bond angles, dihedral angles, or other structural variables such as improper torsions) often form a better basis for performing geometry optimization. This could perhaps be expected given the reasonable accuracy of the molecular mechanics method, which expresses potential energy in terms of relatively simple functions of such coordinates. Here too, though, there is a complication: a 'natural' set of internal coordinates will typically be over-determined, i.e. it will contain more than the $3N-6$ coordinates needed to uniquely identify the structure. For this reason, natural sets of internal coordinates are often described as being *redundant*. When applying a geometry update step, either in a line search or a Newton step, this causes a problem because the combination of changes suggested by the algorithm for each of the internal coordinates will typically not correspond to any possible set of Cartesian coordinates. Consider a three-membered ring ABC, in which one wishes to change each of the angles ABC, BCA, and CAB—only changes that preserve the sum of these angles as 180° are actually feasible, yet a predicted structure update will suggest changes to each of these angles independently. This requires an additional procedure during optimization, to identify the change in coordinates that most closely satisfies all the ones predicted by the algorithm while also being consistent. There is also an additional step involved in converting the gradient from Cartesian coordinates to internal coordinates. Despite these additional complications, optimization in the space of redundant internal coordinates is usually preferred, because it is more efficient in terms of the number of potential energy and gradient evaluations needed to reach the minimum within a given convergence threshold.

Cartesian and internal coordinates for tetrahydrofuran (THF)

THF, C_4H_8O (whose structure is shown in the margin), comprises thirteen atoms and hence its structure can be uniquely defined in terms of thirty-nine Cartesian coordinates. A typical choice of *internal* coordinate system would include eight C–H, three C–C and two C–O bond lengths, six bond angles for each of the four carbon atoms (e.g. H_{A1}–C_A–H_{A2}, H_{A1}–C_A–O, O–C_A–C_B), and one bond angle for the oxygen atom, and a total of thirty-three dihedral angles (nine around each of the C–C bonds, e.g. for the C_A–C_B bond, H_{A1}–C_A–C_B–C_C or H_{A1}–C_A–C_B–H_{B1}, and three around each C–O bond, e.g. H_{D2}–C_D–O–C_A). This makes a total of 71 internal coordinates, many more than are needed to define the structure. Nevertheless, this *redundant* set of coordinates is typically used in quantum chemical studies to describe the structure, as it leads to efficient geometry optimization.

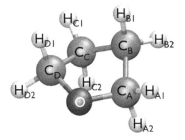

Geometry optimization using a quantum mechanical method

The graph in Figure 5.11 helps to illustrate the outcome of a typical geometry optimization for a small molecule using a quantum mechanical method. The molecule studied was THF, and the geometry was optimized using the Hartree-Fock method, with the 6-31G basis set. Two different optimizations were performed. For the first optimization, the starting structure was obtained by using a graphical program to draw the molecule and to refine the structure by geometry optimization using a simple molecular mechanics method built into the graphics program. The resulting structure—in the form of Cartesian coordinates—was copied and used as input for the quantum chemical code. This program first read the structure, and based on atom-atom distances and knowledge of typical bond lengths, identified bonded pairs of atoms, and used this in turn to choose a set of redundant internal coordinates—in fact the exact set referred to in the previous box. Then the Hartree-Fock equations were solved for the initial structure, returning an energy of −230.859667 hartree. The gradient of the energy with respect to each of the 39 Cartesian coordinates was then evaluated, and converted so as to be expressed in terms of the internal coordinates. The average of the gradient elements with respect to all the internal coordinates (expressed as the square root of the average of the squares of the elements, i.e. the root mean square or RMS gradient) was of 0.0101 (the gradient elements are expressed in hartree per bohr for atom-atom distances and in hartree per radian for angles). As this was above the convergence threshold, a second step was needed. In fact, a total of seven steps were needed, after which the energy (V_1 in Figure 5.11) had dropped to −230.874811 hartree, and the RMS gradient (G_1) had dropped to 0.000026, below the threshold.

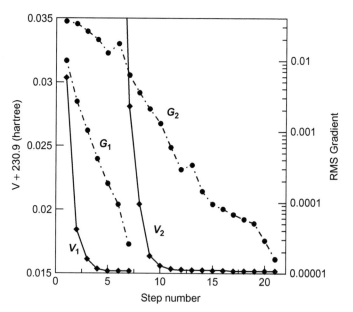

Figure 5.11 Outcome of two HF/6-31G geometry optimizations for THF, using redundant internal coordinates and a form of the BFGS method. Energies V_1 and V_2 at each optimization step are shown as black diamonds and refer to the left-hand scale (V_2 is off the scale for the initial points); RMS gradients G_1 and G_2 are shown as black dots and refer to the logarithmic right-hand scale.

(continued...)

The second optimization followed the same procedure, but in this case, the starting structure was modified by randomly altering the Cartesian coordinates by a few tenths of an Å. The energy (V_2) therefore starts from a higher point, at −230.268472 hartree, and takes more steps to reach convergence. The gradient (G_2) is also larger at the initial point. The optimization takes twenty-one steps. Because the starting point was different, and the optimization terminates once some convergence threshold is met, the final energy obtained is not necessarily identical to that obtained in the first optimization, but in this case the final energies obtained differ only in the eighth digit after the decimal. The structures obtained are nearly identical also, with bond lengths within 0.0001 Å, and bond angles and dihedrals within 0.1°. This shows that the convergence thresholds chosen were low enough to yield very comparable energies and structures, with the error due to the threshold much smaller than the error due to the underlying quantum chemical method. The additional computational effort in the second case illustrates the importance of starting a geometry optimization from as good a guess of the final structure as possible.

THF is a small molecule, with a fairly rigid structure. This means that the optimization proceeds mostly smoothly in both cases, notwithstanding the poorer starting point in the second calculation. Both the energy and the RMS gradient drop at almost every step. For larger molecules, especially those with many flexible dihedral angles or other 'floppy' degrees of freedom, optimization can take many steps, and the approach to convergence can sometimes involve multiple steps where neither the energy nor the gradient improves much. As a rule of thumb, a system with n atoms will take of the order of n steps to optimize using the very efficient second-order optimization algorithms used in contemporary quantum chemistry codes.

The algorithms used in geometry optimization converge to a *local* minimum, corresponding to the same 'basin' in the potential energy surface as the starting point, as shown schematically in Figure 5.12. To take an example, geometry optimization of *n*-butane could lead to the *anti*- or *gauche*-conformers depending on the initial structure. In principle, one might even end up obtaining the minimum corresponding to isobutane (or to butene + H_2, ...) depending on the initial structure used. It is important

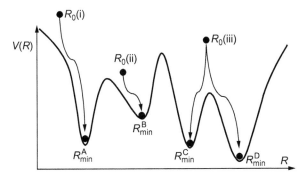

Figure 5.12 Schematic representation of a potential energy surface $V(R)$, with more than one local minimum. Geometry optimization initiated from different structures $R_0(i)$, $R_0(ii)$, and $R_0(iii)$ leads to different local minima R_{min}^{A-D}, usually the one 'closest' to the starting point. For the initial point $R_0(iii)$ with two nearby minima R_{min}^{C} and R_{min}^{D}, optimization could lead to either of them depending on the algorithm used.

to remember that for systems that can exist as multiple conformers or indeed isomers, geometry optimization is not guaranteed to find the overall or **global** minimum. This is a frequent cause of error in computational studies, since the energy and other properties of a local minimum may be rather different from those of the global minimum, which is usually what is relevant in experiments.

Separate methods are needed to locate the *global* minimum (R_{min}^D in Figure 5.12). Such techniques go beyond the scope of the book, but briefly, they effectively rely on exhaustively carrying out standard optimization, starting from different initial structures. This same procedure can be carried out manually for small- to medium-sized systems, with chemical intuition being used to generate the different initial structures.

Most quantum chemical methods yield reasonably accurate optimum structures. Table 5.1 below shows a few simple molecules, with a few experimental values for particular bond lengths, angles, and dihedral angles using different levels of theory. As can be seen, the agreement is fairly good even with the fairly inaccurate Hartree–Fock method. Chapters 2 and 3 dealt in some detail with the fact that all methods yield incorrect overall energies. In order for it also to be the case that structures are predicted incorrectly, then the errors in the potential energy must be different for different structures. This is indeed the case—but near minima on the potential energy surface, the errors are fairly constant, hence the fact that structures are predicted quite reliably.

Table 5.1 also illustrates an interesting point: for some structural features, computation can be *more* accurate than some experiments. Obtaining an accurate value for the bond angle and dihedral angle in hydrogen peroxide from experiment is not trivial, and the value reported in the online database referred to in Table 5.1, based on a microwave spectroscopy experiment and published originally in 1962, is apparently incorrect. More recent experiments suggest a structure much closer to the CCSD(T) values shown in Table 5.1.

The experimental values in Table 5.1 are from gas-phase experiments. More often, when comparing computed structures to experiment, the quantum chemical calculations are performed in vacuum, whereas the experimental method is X-ray crystallography. The crystal environment can somewhat perturb the structure, so that the comparison is not on a like-for-like basis. Changes in structure between vacuum and crystals occur most readily for degrees of freedom along which the energy only varies

Table 5.1 Calculated and experimental bond lengths (r/Å), bond angles (α/°), and dihedral angles (d/°) at the HF, MP2, and CCSD(T) levels of theory using the cc-pVTZ or cc-pVQZ basis sets. Experimental values from the computational chemistry benchmark database (http://cccbdb.nist.gov/, accessed 29 June 2017).

Molecule		HF/VQZ	MP2/VQZ	CCSD(T)/VTZ	CCSD(T)/VQZ	Exp
CH_4	r_{CH}	1.082	1.084	1.089	1.088	1.087
H_2O	r_{OH}	0.940	0.958	0.959	0.958	0.958
	α	106.2	104.0	103.6	104.1	104.5
HOOH	r_{OH}	0.941	0.963	0.964	0.963	0.950
	r_{OO}	1.385	1.446	1.458	1.452	1.475
	α	103.1	99.7	99.5	99.9	94.8
	d	111.3	112.7	113.9	112.5	119.8

modestly, or 'soft' degrees of freedom. The optimum torsion angles around single bonds can be affected in this way (for example, in the crystal structure of hydrogen peroxide, the dihedral angle is much smaller, at 90°), while most bond lengths and angles are less perturbed by crystal packing or solvent interaction. Even gas phase structures measured by experiment are slightly different from the structure at the minimum of the exact potential energy surface: effects such as zero-point energy or additional rotational or vibrational energy introduce small distortions.

5.5 Geometry optimization using molecular mechanics

Broadly speaking, geometry optimization using MM methods can be carried out for much larger systems than are accessible with quantum chemical methods due to the lower computational effort required to evaluate the potential energy and the gradient. In principle, the same methods can be used for the optimization. There are, however, some subtle differences when optimizing large structures, because the much larger number of degrees of freedom means that the number of numerical operations needed to perform the optimization can begin to rival that for the calculation of the energy and gradient in terms of their relative computational demands. *Storage* of the Hessian matrix in the computer's memory can also be challenging. As a result, while second-order methods based on equation (5.7) are used also for MM geometry optimization, the exact algorithms used are usually different, with efficiency now needing to be tempered by computational cost of the operations needed for the algorithm, and by issues related to storage. Conversion from Cartesian to internal coordinates also becomes more challenging, so that optimization is usually performed using Cartesian coordinates.

Geometry optimization using the TIP3P water model

Geometry optimization for a system of 1600 water molecules described using the TIP3P water model (modified to describe also the intramolecular O–H bond stretching and bending terms), occupying a cubic box with length 36.342 Å, is a way to understand some of the features of geometry optimization for large systems. The initial structure was selected randomly from a molecular dynamics simulation (see Chapter 6), and was more than 1,000 kcal mol^{-1} above the local minimum finally reached. This system has 4,800 atoms or 14,400 degrees of freedom, and is thereby at the low end compared to typical MM system sizes treated in research projects (typically 1,000 to 1,000,000 atoms, though larger and smaller systems are studied also). Storing the full Hessian (or inverse Hessian) for this system becomes challenging, since it is a $14,400 \times 14,400$ matrix. Even taking into account that this matrix is symmetric, so only half of it (plus the diagonal) needs to be stored, this still represents a storage requirement of over 100 million real numbers. Carrying out operations on such a large matrix is also very demanding. Accordingly, for this case the full Hessian was not used, and instead two approximate second-order methods were used for the optimization. The first algorithm was selected for efficiency, and was able to carry out each structural update quite quickly, though after over 1,000 steps, the energy was still over 5 kcal mol^{-1} above the final minimum, and the RMS gradient was still of about 0.1 kcal mol^{-1} Å^{-1}. A second, more accurate but slower, algorithm was selected to reach convergence to within better than 0.001 kcal mol^{-1} in energy terms, and to an RMS gradient below 0.001 kcal mol^{-1} Å^{-1}, in just a few tens of steps.

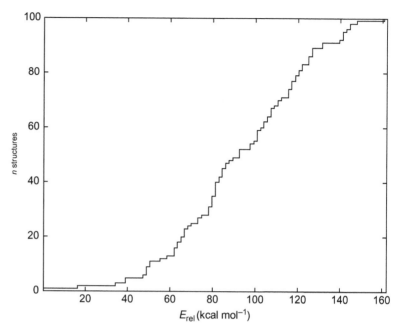

Figure 5.13 Summary of the outcome of 100 separate geometry optimizations of a periodic box containing 1,600 water molecules, starting from 100 different initial structures. The cumulative number of structures with energies equal to or lower than a given energy E_{rel} relative to the lowest minimum found is shown as a histogram (e.g. roughly 30 structures lie within 80 kcal mol^{-1} of the lowest minimum found).

Geometry optimization is regularly used as part of studies of large systems with molecular mechanics methods. However, unlike for small systems, the resulting energies are seldom used, with the optimization instead being performed as one step in the preparation of the system for molecular dynamics studies (see Chapter 6). The reason for this difference is that for large systems, the conformational complexity is very large, far greater than that shown in the cartoon of Figure 5.12, such that it becomes impossible in practical terms to locate the global minimum or to control which local minimum is obtained. This is illustrated in Figure 5.13, where 100 separate geometry optimizations of the structure of a box of water molecules were carried out, all using the same forcefield and procedure as was described in the preceding box, but starting from slightly different structures (taken from a molecular dynamics simulation, see section 6.3). As it happens, all the final structures and energies obtained were different. Relative to the overall energy of −20,773.0362 kcal mol^{-1} for the lowest-energy structure found, the other 99 structures span a range of over 160 kcal mol^{-1}.

5.6 Properties of optimized structures: vibrational frequencies

Geometry optimization yields a structure that is a minimum of the potential energy surface, which is a point where the gradient—or first derivative—of the potential energy with respect to displacements of the atoms is zero. The *second* derivative of the energy is not usually zero at the minimum. This second derivative is the Hessian matrix that was already mentioned, and it can be very useful to calculate it, as it provides information concerning infrared spectra and other aspects of vibrational spectroscopy. The Hessian matrix, after weighting to take into account the masses of the different atoms, can be used to generate a set of eigenvectors and eigenvalues that give a reasonably accurate description of the vibrational energy states of the system. The eigenvectors are the vibrational 'normal modes' of the system, that is, a set of privileged coordinates that can be used to describe distortions of the system. These normal modes or distortions are such that they do not couple to other distortions—motion initiated in one normal mode does not then transfer to another normal mode (more precisely, it only transfers when taking account of the terms in the potential energy due to higher derivatives).

The corresponding eigenvalues measure the stiffness of the potential energy surface along the corresponding distortion trajectory given by a particular eigenvector. This is like a multi-dimensional generalization of the force constant for diatomic molecules, and like force constants, the eigenvalues are related to the vibrational frequencies for the molecule through a square root relationship. At a saddlepoint on the potential energy surface, one of the eigenvalues will be negative (leading to an imaginary frequency), with the corresponding eigenvector corresponding to the reaction coordinate leading away from this TS structure. Minima on the potential energy surface have only positive eigenvalues, so the number of negative eigenvalues can be used to check whether a structure is a minimum or a saddlepoint (or, if it has more than one negative eigenvalue, a higher-order saddlepoint). Note that frequency analysis is only strictly meaningful at a minimum or saddlepoint (a stationary point).

As mentioned above, quantum chemical (and molecular mechanical) methods yield errors for the potential energy that are not constant across the range of structures. This yields small errors in predicted molecular structures, and also leads to errors in the Hessian matrix elements and hence in the vibrational eigenvalues. The errors on the eigenvalues are somewhat larger than the errors in bond lengths or angles. As for molecular structures, some of the error compared to experiment is due to the quantum chemical method, and some can be due to the fact that infrared spectra, for example, are typically measured in solution or in a crystal—with the environment perturbing the vibrational frequencies of the isolated molecules. There is a third source of error to take into account: observed vibrational frequencies depend on the energy difference between the ground, zero-point level of a particular vibrational mode of a molecule, and a vibrationally excited state. The energy of these states depends not only on the second derivative of the energy at the position of the minimum, but also on the detailed shape of the potential energy surface further from the minimum. This is fairly well predicted by the Hessian matrix—which contributes the first non-trivial term to a Taylor expansion of the potential energy around the minimum. However, higher-order terms have a detectable effect on the observed frequencies. Using only the second-order terms is known as the *harmonic approximation*. Some examples showing the sort of level of agreement that can be expected between experiment and theory are shown in Table 5.2.

Calculation of vibrational frequencies is often used to generate various corrections to the potential energy of a system, as will be discussed in Chapter 7. The overall zero-point energy due to the Heisenberg uncertainty principle is one example.

Table 5.2 Calculated and experimental harmonic frequencies (in cm^{-1}) for three diatomic molecules, at the Hartree–Fock (HF), MP2, and CCSD(T) levels of theory, using various cc-pVXZ basis sets (X = D: VDZ, X = T: VTZ, X = Q: VQZ). Experimental values from the NIST webbook, http://webbook.nist.gov/chemistry/form-ser/.

Molecule	HF/VQZ	MP2/VQZ	CCSD(T)/ VDZ	CCSD(T)/ VTZ	CCSD(T)/ VQZ	Exp.
H_2	4582	4520	4382	4409	4403	4401
HF	4478	4159	4150	4177	4162	4138
CO	2427	2128	2143	2153	2164	2170

5.7 Transition states and reaction paths

All the above discussion has concentrated on describing geometry optimization for minima. When studying reactivity, it is also interesting to be able to locate transition states, or saddlepoints, on the potential energy surfaces. This can also be done using geometry optimization methods, but is more complicated, because in this case the 'steepest descent' philosophy is no longer applicable: a saddlepoint is a (local) *maximum* in one direction, and a minimum in all the others, as shown e.g. in Figures 5.2, 5.3, and 5.6. Following the gradient 'downhill' will, therefore, almost always lead away from the saddlepoint. The exception is when the TS structure differs in its symmetry properties from the structures lying along the reaction path on either side of it. For example, the TS for the simple exchange reaction $H + H_2 \rightarrow H_2 + H$ is symmetric, with $r(H_A–H_B) = r(H_B–H_C)$, whereas on the reactant side, for example, $r(H_A–H_B) > r(H_B–H_C)$ and the symmetry is broken. Most geometry optimization methods preserve symmetry, so if one starts a geometry optimization for the H_3 system from an initial structure in which the two H–H distances are exactly equal, the structure obtained is the TS. Except for these (rare but sometimes useful) cases, TS optimization requires additional care, and specialized methods for carrying it out are used.

The simplest method is to use a second-order method using a Newton step (equation 5.7), using an inverse Hessian that has either been computed exactly, or at least reasonably accurately in the key direction corresponding to the reaction coordinate. When applied near a potential energy minimum, the values in the Hessian matrix correspond to positive curvature of the potential energy surface in all directions. Therefore the structure change ΔR that will be produced by equation (5.7) will be broadly opposite in direction to the gradient. Since the gradient points 'uphill', ΔR is downhill. If, however, one is near a saddlepoint, then the surface will be concave in one direction, i.e. **H** will have one negative eigenvalue—and **H**$^{-1}$ will also behave like a negative number for the corresponding direction or eigenvector. This will lead the Newton step to be *uphill* in that direction, as is needed to move towards the saddlepoint. This procedure can be a very effective one for locating a TS, but it requires a very good initial guess of the TS structure, and a very accurate estimate of the Hessian at this initial guess structure.

A second important method for locating TSs is to provide, as well as an initial guess structure, also an initial guess of the direction of the reaction coordinate. With this information, the optimization algorithm can force itself to make a step that is uphill along the reaction coordinate, even if the estimated Hessian is not accurate enough to reflect the concave nature of the potential energy surface in that direction, and/

or if the initial structure does not even lie in the fairly small region where the surface is indeed concave. A convenient way to provide both the structure and the guessed reaction coordinate is to provide *two* structures that are believed to lie at different points near the minimum energy path from reactants to products: one is more 'reactant-like', the other more 'product-like'. The difference between the two structures can then be taken as the search direction, and the average of the two is used as the initial structure for the search. A version of this method that is available in many quantum chemistry codes is called the quasi-synchronous transit (QST) method. In a variant of this method, one can provide *three* input structures: one reactant-like and one product-like, with the difference between two of them defining the initial search direction, and the third being (hopefully!) TS-like.

Applying the quasi-synchronous transit method

An example of application of the quasi-synchronous transit method is the Diels–Alder reaction, in the simplest case of ethene C_2H_4 with 1,3-butadiene $H_2C=CH-CH=CH_2$. This reaction converts two separate molecules into one. The two input structures for the QST method cannot, therefore, both be structures of stable species. Instead, one can use the optimum structure for cyclohexene, together with a structure in which butadiene has started to approach ethene with the appropriate orientation for reaction. This requires a cisoid conformation of butadiene, with the ethene at an appropriate pre-reactive position below the butadiene plane, and with appropriate C–C distances, here chosen to be slightly smaller than the C–C van der Waals radius of 3.4 Å. This structure was accordingly built and loosely optimized, while holding both C–C distances at 3.0 Å. Together with the cyclohexene structure, this pre-reactive structure was used as input for the QST2 approach (taking care that the atoms in both structures were ordered in the same way, so as not to cause confusion in the algorithm that analyses their differences in positions). The TS search algorithm 'knows' that the reaction coordinate lies along the direction corresponding to the difference in these two structures, and rapidly converges to a TS structure, shown in Figure 5.14 along with the pre-reactive and product structures. At the B3LYP DFT level, with the 6-31G(d) basis set, the separate reactants butadiene (in cisoid form) and ethene lie at –155.98649 and –78.58746 hartree, respectively, or –234.57395 hartree considered together. Their pre-reactive approach structure lies 13 kJ mol^{-1} higher at –234.56897 hartree, the TS lies 79 kJ mol^{-1} above reactants at –234.54390 hartree and cyclohexene lies 195 kJ mol^{-1} below reactants at –234.64829 hartree.

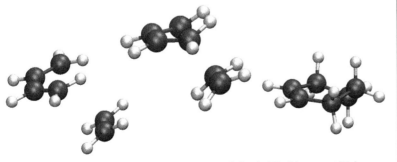

Figure 5.14 Structures generated during a QST search for the TS of the parent Diels–Alder reaction: pre-reactive complex (left), TS (middle), cyclohexene (right).

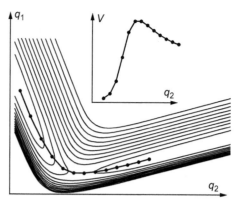

Figure 5.15 Approximate TS location using a relaxed scan on a model potential energy surface. Successive geometry optimizations are carried out using different fixed values of the coordinate q_2, with the other coordinates (in this simple example, there is only one, q_1) being optimized. The dots show the positions of these optimized structures, with the inset showing the energy along the scan.

In some cases the procedures described above are not sufficient to locate TSs. In such cases, it can be helpful to first carry out a so-called 'relaxed scan' of the potential energy surface. This requires a set of calculations, in each of which the structure is optimized (or relaxed), while holding one of the coordinates, for example a bond length, at a given fixed value. If this coordinate changes smoothly along the reaction path, then the optimized structures obtained for the different fixed values will lie along the reaction path, and one of them will even be near the TS. A plot of the energy V of the optimized structures as a function of the constrained bond length value will go through a maximum near the TS (see Figure 5.15 for an example). This structure (or two structures taken from the scan) may then be used as a starting point for optimizing the TS structure exactly using the methods described on previous pages. This method does not work well when small changes in q_2 lead to large moves along the reaction coordinate. The simple example of Figure 5.15 uses effectively the same potential energy surface as in Figure 5.6, but modified so that the coordinate q_2 is more or less parallel to the reaction coordinate throughout the reactive region. A relaxed scan along q_2 for the surface of Figure 5.6 would be less effective.

Another way of describing the problem with relaxed scans is to observe that the set of structures obtained will not necessarily be equally spaced along the reaction path. Even though each pair of consecutive structures may be chosen to differ by the same amount for the constrained distance q_2, the other coordinates are optimized at each step and the difference between their optimized values for adjacent points on the path can vary, as shown in Figure 5.15. In unfavourable cases, a very small change in the scanned coordinate can lead to a very large structural change, which can for example completely bypass the TS. Careful choice of the scanned coordinate is mandatory, and an optimal choice may not be possible if the reaction involves complicated structural changes.

Other methods exist to locate whole reaction paths without needing to specify one particular coordinate in advance. The 'nudged elastic band' (or NEB) method is one such popular technique. To use this method, one first chooses a fixed starting structure

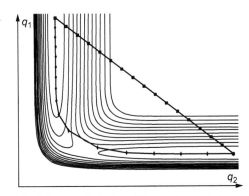

Figure 5.16 Nudged elastic band optimization of the reaction path for a reactive potential energy surface.

R_{init}, a fixed final structure R_{fin}, and N intermediate structures $R_1^0, R_2^0, ... R_N^0$. The coordinates for these intermediate structures R_i are then optimized so as to minimize (a) the potential energy for each of them, and (b) the distance between each pair of adjacent structures, $|R_i - R_{i+1}|$ (including the distance from R_{init} to R_1 and from R_N to R_{final}). The result is an approximate reaction path formed by the sequence of points R_{init}, R_1, ... R_N, R_{final}. There are several variants of the NEB method, which differ e.g. in the form of the expression used for the distance term $|R_i - R_{i+1}|$. An example of the application of the nudged elastic band method is shown in Figure 5.16. The NEB method is frequently used for studying reactions in very large systems, where other algorithms typically used for locating TSs become unwieldy.

Once a TS has been located, it is also possible to work out the reaction path using a variety of methods. One very commonly used approach generates the so-called 'intrinsic reaction path', which follows the 'intrinsic reaction coordinate'. This is defined as the path that descends most steeply from the TS, in terms of coordinates that are weighted by the masses of the atoms. A final note on TS and reaction path optimization is to point out that, as for minima, there may be many conformational variants, and typical optimization methods only give the 'local' minimum in the space of transition state structures.

5.8 Further reading

- Geometry Optimization. H. B. Schlegel, *WIREs Computational Molecular Science*, 2011, **1**, 790–809. This review by one of the world's experts on geometry optimization using quantum chemical methods gives an overview of modern methods.

- *Numerical Recipes: The Art of Scientific Computing*, 3rd Edition. William H. Press, Saul A. Teukolsky, W. T. Vetterling, and Brian P. Flannery, Cambridge University Press, Cambridge, 2007. Geometry optimization is just one example of a general numerical problem for which computers are often used. For readers wishing to understand the mathematical and computational background of optimization, this textbook provides a useful entry point.

- *Energy Landscapes: Applications to Clusters, Biomolecules and Glasses.* David J. Wales, Cambridge University Press, Cambridge, 2003. This book describes the many features of potential energy surfaces for chemical systems, and the links between these features and the physical behaviour of the corresponding system.

5.9 Exercises

5.1   Using a quantum chemical code and a simple level of theory such as HF/6-31G, optimize the structure of butane starting from a number of different structures. From visual inspection of the final structures and energies, identify which initial structures lead to the same local minimum, and which yield different minima. Locate the minima corresponding to gauche, anti, and isobutane. Also, see how many optimization steps are needed, depending on the quality of the input structure.

5.2   Using symmetry, optimize the structure of the TS for the reaction $FH + F \rightarrow F + HF$ using a simple level of theory such as HF/6-31G. Calculate the vibrational frequencies for the TS. You should find that one frequency is 'negative' (actually it is imaginary)—this denotes that curvature of the potential energy surface is negative along the direction of this eigenvector. This is the reaction coordinate.

5.3   Using the QST2 method, locate the TS for the model Diels–Alder reaction, following the procedure described near Figure 5.14.

5.4   Using a molecular mechanics code, carry out geometry optimization for a large system such as a box of water molecules. Identify the keywords that select different approximate second-order optimization methods, and test to see how rapidly the RMS gradient drops.

5.10 Summary

- Geometry optimization is one of the main techniques used to explore potential energy surfaces.
- Efficient optimization methods are based on calculation of the potential energy, its first derivative with respect to the positions of the atoms (the *gradient*), and the second derivative (the Hessian matrix).
- Geometry optimization is frequently used in quantum chemical studies; predicted molecular structures usually agree very well with experiment.
- The curvature of the potential energy surface around the optimum structure can be used to predict vibrational frequencies.
- Geometry optimization for extended systems as studied with molecular mechanics presents new challenges, especially given the very large number of local minima typically present on the potential energy surface.
- As well as minima, optimization can also be used to locate transition states and reaction paths.

6 Dynamics Methods

6.1 Introduction

Geometry optimization is a very useful way to explore the properties of a potential energy surface. Information about the minimum energy structures and TSs does not however tell you exactly how the chemical system described by the energy surface will behave: how the atoms making up the system will be distributed in space, and how their positions will evolve with time. This is instead described by the *chemical dynamics* of the system.

Chemical dynamics can in principle be described in terms of the Schrödinger equation, see Chapter 2. Both the time-dependent and the time-independent versions of the Schrödinger equation can be used to predict the dynamics of molecular systems. These *quantum* dynamical methods are beyond the scope of this book, though it can be noted that the methods to compute vibrational eigenvectors and frequencies described in section 5.6 are one example of a simple approximate quantum treatment of dynamics.

While a quantum mechanical treatment of the motions of *electrons* is almost always necessary, due to their very small mass, atoms are sufficiently heavy that a classical mechanical description of their motions is often adequate. In this chapter, we will focus on this classical description and its use for describing chemical behaviour.

6.2 Newton's laws of motion

To describe atomic motions using classical mechanics, the starting point is the well-known Newtonian law of motion, equation (6.1):

$$F = ma \tag{6.1}$$

In equation (6.1), F is the force acting on an atom, m is its mass, and a is the acceleration that it experiences. The same equation also applies for a collection of N atoms, where F is now a $3N$-dimensional vector (or an $N \times 3$ matrix) collecting the force applying to each atom, m is the masses of the atoms, and a is a vector (or matrix) collecting all their acceleration terms. The vector F is directly linked to the potential energy surface $V(R)$: it is simply the negative of the gradient, equation (6.2):

$$F = \frac{-\partial V(R)}{\partial R} \tag{6.2}$$

For some very simple potential energy surfaces, Newton's equation of motion can be solved exactly with pencil and paper. This is the case for a harmonic oscillator, where the particle follows a sinusoidal trajectory with an amplitude that depends on the initial conditions, i.e. the initial position and velocity. For most potential energy surfaces, though, Newton's equations can only be solved (or integrated) in an approximate numerical way: for each atom, the initial position R_0 and velocity $v(0) = dR_0/dt$ are assigned, then Newton's equation is used to predict the new positions and velocities after a short time interval Δt. One simple way to do this is to start from the definition of the velocity $v(t)$ and the acceleration $a(t)$ at each point in time as derivatives of the position and the velocity, respectively, equation (6.3):

$$\begin{cases} v(t) = \dfrac{\partial R(t)}{\partial t} = \lim_{\Delta t \to 0} \dfrac{R(t+\Delta t) - R(t)}{\Delta t} \\ a(t) = \dfrac{\partial^2 R(t)}{\partial t^2} = \dfrac{\partial v(t)}{\partial t} = \lim_{\Delta t \to 0} \dfrac{v(t+\Delta t) - v(t)}{\Delta t} = \dfrac{1}{m} \times \dfrac{-\partial V(R)}{\partial R} \end{cases} \tag{6.3}$$

Instead of going to an infinitesimally small Δt, one can consider a finite step in time, and rearrange, to obtain equations giving the position and velocity:

$$\begin{cases} R(t+\Delta t) = R(t) + v(t)\Delta t \\ v(t+\Delta t) = v(t) + a(t)\Delta t \end{cases} \tag{6.4}$$

These expressions in equation (6.4)—called the Euler equations—are intuitively appealing but in practice are not very useful, because both the velocity and the acceleration vary during the course of the time-step, leading to errors in the predicted changes in position and velocity. By choosing a small enough time-step Δt, the error incurred over one step can be kept small, but this increases the number of time-steps n_{step} needed in order to reach an overall simulated time τ ($n_{step} = \tau/\Delta t$), and the cumulative error over many time-steps remains too large also. More sophisticated procedures for integration have been developed which yield much smaller errors and allow much larger time-steps. One example of such an integration method is called the *Velocity Verlet method*, a member of a popular family of techniques first used for simulating molecular dynamics by the French physicist Loup Verlet. The equations (6.4) above are modified to equations (6.5):

$$\begin{cases} R(t+\Delta t) = R(t) + v(t)\Delta t + \dfrac{1}{2}a(t)\Delta t^2 \\ v(t+\Delta t) = v(t) + \dfrac{1}{2}\{a(t) + a(t+\Delta t)\}\Delta t \end{cases} \tag{6.5}$$

To apply the Velocity Verlet method, initial positions $R(0)$ and velocities $v(0)$ are chosen, and the energy and gradient of the energy are computed. The gradient yields the acceleration vector at the initial position $a(0)$ using Newton's equation of motion (6.1), and then the first part of equation (6.5) can be used to predict the new position $R(0+\Delta t)$. The energy and gradient at this new position can then be calculated, yielding $a(\Delta t)$ which can be used together with the second part of equation (6.5) to predict the updated velocity $v(\Delta t)$. The whole procedure is then repeated n_{step} times.

Calculation of a classical dynamics trajectory using the Velocity Verlet method or related approaches can be useful in a number of contexts. One example is the study of the detailed *dynamics* of chemical reactions. Consider the reaction $A + BC \to AB + C$ that

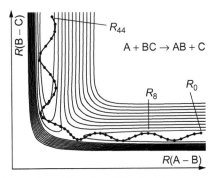

Figure 6.1 Classical trajectory from integration of Newton's law on a reactive potential energy surface. Starting from position R_0 corresponding to reactants A + BC, the trajectory moves in 44 steps to AB + C. Such trajectories can be used to predict the conversion of the collision energy into vibrational energy.

could be described (Figure 6.1) by the simple potential energy surface example already shown as Figures 5.6, 5.15, and 5.16. You might wish to predict whether this reaction proceeds more readily when the reactants (A and BC, bottom right of Figure 6.1) approach each other and collide with a high relative energy, or whether the same amount of energy imparted to the B–C vibrational motion is more likely to lead to reaction. Considering the products AB + C, you might also wish to know how likely they are to contain a certain amount of energy in the form of relative velocity as they move apart, or as vibrational energy of the A–B molecule. Figure 6.1 shows an example of a trajectory calculated using initial coordinates R_0 and with the initial velocities corresponding to the two partners moving towards each other, with a small amount of vibrational energy for BC (note the variation of $R(BC)$ in the initial phase of the reaction). In this example, after $n_{step} = 44$ steps, the trajectory has reached the product region, and it can be seen that the products are predicted to be undergoing fairly strong vibration.

6.3 Molecular dynamics simulations

A more common use for calculation of classical trajectories is for the study of the behaviour of large, complicated systems, where the large number of different local minima on the potential energy surface makes the use of geometry optimizations unenlightening for predicting most interesting properties of the system (see Figure 5.13). In this context, the operation of integrating of Newton's law of motion is usually referred to as carrying out a *molecular dynamics* (or MD) simulation. The generated trajectories typically visit many of the local minima on the potential energy surface (more discussion on this point is deferred to Chapter 7), and suitable averages over the positions and/or velocities at each time-step can be computed, and compared to experiment in a way that properties of a single potential energy minimum cannot.

Carrying out an MD simulation is a procedure that typically requires a number of different parts, as summarized in Figure 6.2. Although *selection of the model system* to be studied is to a large extent imposed by the nature of the chemistry, choices need to be made with care, since the larger the system, the more demanding the calculations. Typical molecular mechanics forcefields include at least one energy term per pair of

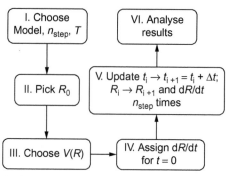

Figure 6.2 Steps involved in setting up and running a molecular dynamics simulation.

atoms, given the presence of the non-bonded interactions, as discussed in Chapter 4. This would suggest that evaluating the potential energy should require a number of operations increasing with the square of the number of atoms in the system. The use of cut-offs and related procedures can reduce this to near-linear scaling, but very large systems are still demanding, so there is a benefit to making the model as small as possible. One way to do this is to identify the largest length scale that contributes in a significant way to the phenomenon of interest, and to use this to define the minimal model size. Very often physical intuition can be used as a guide, or you can run some exploratory calculations with different sizes.

Second, you need to generate an *initial structure* for the atoms in the system. In QM studies of small or medium-sized molecular systems, this is frequently done manually by 'building' the system using text input or a graphical interface. In some simple cases, such as a liquid comprised of simple molecules (e.g. water, or the even simpler liquid rare gases), the multiple replicas of the molecule being studied can simply be placed randomly in a simulation box with periodic boundary conditions. Likewise simple crystalline solids can be built up from knowledge of the features of the unit cell. For more complex systems such as biomolecules, however, input from an experimental structure is often needed. Proteins have very detailed secondary, tertiary, and, in some cases, quaternary structure that cannot simply be assigned based on knowing the amino acid sequence. Even with systems such as biomolecules where experimental input is used for the positions of some of the atoms, additional work is needed to assign coordinates to *all* the atoms in the model. This includes the hydrogen atoms (which are usually missing in X-ray crystal structures), the solvent molecules, counter-ions, and so on.

The third step is to choose the method used to calculate the potential energy. This can be an *ab initio* quantum chemical, DFT, molecular mechanics, or hybrid quantum mechanical and molecular mechanical (see Chapter 8) method. The most common choice is molecular mechanics. In that case, one needs to specify all the forcefield terms to be used: bond stretching, angle bending, dihedral torsions, partial charges, van der Waals interaction terms, and so on (see Chapter 4). In simple cases, the system to be treated will only contain bonding patterns that are sufficiently analogous to ones that have already been successfully modelled by others. Then you simply need to choose a forcefield family, to define the bonding connectivity of the atoms in the system and to assign each atom to a given 'atom type'. If new atom types are involved, though, it will be necessary to provide parameters for them, using the approaches described in Chapter 4.

In the case of DFT evaluation of the potential energy, a special procedure can be used whereby the solution of the Kohn–Sham equations is *coupled* to dynamic equations similar to equation (6.5) yielding a method that is less computationally demanding. This so-called 'Car–Parrinello' method is named after its developers, the Italian physicists Roberto Car and Michele Parrinello.

The fourth step is to initiate the molecular dynamics by assigning initial velocities to all the atoms. As discussed below, the goal in MD is often to model behaviour at a given temperature. Temperature is not a microscopic variable (e.g. it is absent in the Euler and Verlet equations above), but from classical statistical mechanics, one knows that at a given temperature T a system with N atoms will have an overall kinetic energy defined according to equation (6.6) based on the equipartition principle:

$$\text{K.E.} = \sum_{i=1}^{N} \frac{m_i |v_i|^2}{2} = \frac{3}{2} N k_B T \qquad (6.6)$$

By choosing 'random' velocities for each atom, then scaling each velocity by a constant, it is possible to ensure that the overall kinetic energy matches that given by equation (6.6).

In many cases, the initial structure chosen in the second step lies much higher in potential energy than the minimum of the potential energy surface expression. This occurs for example if some atoms have been positioned in a random way, and they happen to be very close to other atoms, and experiencing strongly repulsive interactions. Structures from X-ray crystallography also frequently have such 'bad contacts'. In these cases, it can be helpful to carry out some geometry optimization prior to initiating the dynamics.

The fifth part of the procedure is to propagate the trajectory for the required number of time-steps n_{step}. This is usually by far the most computationally demanding part of the whole process. To integrate the trajectory, one needs to choose the time-step Δt to be used. This needs to be as long as possible in order to reach a long total trajectory time $\Delta t \times n_{step}$ with a small number of steps n_{step}, but small enough to ensure that the integration is accurate. This second aspect requires Δt to be shorter than the period of the fastest motion in the system. In most cases, this fastest motion is vibration of X–H bonds, with typical vibrational frequencies of 3,000 to 4,000 cm^{-1} and vibrational periods of roughly 10×10^{-15} s, or 10 fs. Depending on the type of simulation, Δt is typically chosen somewhere between 0.1 and 1 fs. Special techniques with names such as 'SHAKE' or 'RATTLE' can be used to 'freeze' all X–H bond lengths, in which case longer time-steps of 2 fs or more can be applied.

You also need to choose the statistical mechanical framework to apply within the simulation. The most obvious approach is to simply implement Newton's equations (in the form of Verlet's equations (6.5) or related ones). In the case of a system with periodic boundary conditions, and in the limit of perfect accuracy of the integration scheme, this leads to a trajectory in which the system has the same total energy E, the same number of atoms N, and the same volume V at each time-step. The trajectory is referred to as being run under *NVE* conditions. Sometimes, though, you may wish to obtain insight into the behaviour of the system at a given T. This can be achieved to some extent by choosing the initial velocities to satisfy equation (6.6). However, especially for small systems or for poor choices of initial structure, the overall kinetic energy can drift some distance away from satisfying equation (6.6) during the simulation if no adjustments are made. To counteract this, it is possible to use a so-called *thermostat* to enforce equation (6.6), either by periodically scaling all velocities or with more sophisticated procedures. Trajectories run in this way conserve N, V and T (but E can change)—and are thereby described as using *NVT* conditions. In a way similar to equation (6.6), it is also possible to assign a *pressure* to a given set of atom coordinates and velocities within a periodic simulation system. To simulate behaviour at a given *pressure*, small adjustments to the volume of the periodic box can be made using a *barostat*. Such simulations are then run under *NPT* conditions.

The sixth and final part of the process is analysis of the output of the simulation. For each problem the analysis to be performed will be different. In the example of simulation of water illustrated in the Box, the focus is on examining the distribution of O–O distances present during the simulation. In other cases, as will be discussed in Chapter 7, the emphasis may be on extracting energetic information.

Molecular dynamics simulation of water

In this box, the whole procedure for running an MD simulation to model the mean structure of liquid water is described.

1. The system size is chosen. It is known that water molecules adopt well-defined relative positions due to hydrogen bonding, with O–O distances of around 3 Å. However, unlike in ice, this locally ordered structure does not extend over long distances, with most structure lost for O–O distances $\cong 10$ Å. With periodic boundary conditions, artefacts arise at distances greater than half of the length of the box. So we will choose a box with length 20 Å. We will choose to model a typical density of 0.997 g cm^{-3} or 997 kg m^{-3}, so the mass of water in the box is $(20 \times 10^{-10}$ m$)^3 \times$ 997 kg m$^{-3} = 7.976 \times 10^{-24}$ kg. Based on molar masses of 15.999 and 1.008 amu for oxygen and hydrogen and Avogadro's number of 6.022×10^{23}, the molecular weight of water is 2.9915×10^{-26} kg. The ratio of these masses suggests that our box should contain 266.62 molecules. Since we can only model whole molecules, we choose to model a slightly larger system, with 270 molecules. The box volume is then scaled by 270/266.62, and its length by the cube root of this (1.0042), so the new box length is 20.084 Å.

2. An initial structure for the system is built. First, the structure for one molecule is assigned with r(O–H) = 1 Å and an angle H–O–H of 100° (the precise values here are not important). Then 269 replicas are generated, with random x, y, and z positions within the box, and random orientations in space. Because water is a fairly simple system, the use of a random initial structure is adequate.

3. The potential energy is calculated using molecular mechanics, with the simple TIP3P forcefield for water. The O–H bonds are assigned ideal bond lengths r_0 and force constants $k_{stretch}$ in equation (4.1) of 0.9572 Å and 600 kcal mol^{-1} Å^{-2}, respectively. The bending terms θ_0 and k_{bend} are chosen as 104.52° (or 1.8242 rad) and 75 kcal mol^{-1} rad^{-2}, respectively. The O atoms have partial negative charges of –0.834 au and the H atoms have matching positive charges of 0.417 au, which interact with partial charges on *other* water molecules according to equation (4.6), with the permittivity ε equal to that of vacuum, $\varepsilon_0 = 8.8542 \times 10^{-12}$ F m^{-1} or $1/4\pi\varepsilon_0 = 332.06$ kcal mol^{-1} e^{-2} Å, where e is the charge of an electron. The oxygen atoms interact through the Lennard-Jones potential of equation (4.5) with a well-depth ε of 0.1521 kcal mol^{-1} and a radius σ_{OO}

(continued...)

of 3.15061 Å. No Lennard-Jones interactions are used for the hydrogen atoms. All of these values were assigned following research in the group of theoretical chemist William L. Jorgensen, with the aim of developing a forcefield that reproduces the observed properties of liquid water. In the simulation here, both the coulombic interactions of equation (4.6) and the Lennard-Jones interactions of equation (4.5) are smoothly cut off using a combination of equations (4.8) and (4.9), so that they reach zero at $r = 9$ Å.

4. The initial random position has a high potential energy of over 10^{15} kcal mol^{-1}, due to some of the randomly placed water molecules being very close to one another. Before starting the MD simulation, some geometry optimization is performed, to reduce the root-mean-square gradient to less than 10 kcal mol^{-1} Å^{-1}. This is done in two phases: first, only the intramolecular stretching and bending terms, and the Lennard-Jones terms are used. Then the optimization is repeated using the coulombic terms as well. The reason for doing this is that in the initial structure, some hydrogen atoms are very close indeed to the oxygen atom of another water molecule, and the forcefield used has the artefactual property that there is a minimum corresponding to structures in which an H atom is superimposed on an O atom of a different molecule. This deep minimum is separated from 'normal' structures by a large barrier, but may be the 'closest' minimum upon optimization from a randomly chosen initial structure. Such artefacts at short range occur in several types of forcefield; initial optimization with a modified potential as described here can then be useful to ensure that one reaches the 'normal' part of the potential energy surface. Then velocities are assigned using equation (6.6) and a target temperature of 298 K.

5. The simulation is performed under NVT conditions, using a time-step of 1 fs and a thermostat to keep the temperature close to 298 K. The first part of the simulation involves structures that are different in terms of their energy and hydrogen-bond pattern compared to those visited later in the simulation. This is because the starting point of the simulation, the partially optimized structure from part 4, is not typical of liquid water as it still partly reflects the random way in which it was generated. After an initial period called the *equilibration phase* of the simulation, the system will have reached a typical equilibrium distribution. In total, 50,000 time-steps (or 50 ps) of simulation are performed. On a modern computer, for the size of system considered here, depending on the software program used, the whole simulation should take a few minutes to a few tens of minutes.

6. The simulation is then analysed. While many different analyses can be performed, here the *radial distribution function* $g(r_{OO})$ is calculated. This function measures the average density of oxygen atoms at a given distance r_{OO} from another oxygen atom, divided by the bulk density of oxygen atoms. It can be computed by counting the average number of oxygen atoms that are present in the small volume $4\pi r^2_{OO}dr$ situated at distances between r_{OO} and $r_{OO} + dr$ from a given oxygen atom at each time-step in the simulation, and dividing by the number that would be expected for the bulk. For small r_{OO} this function equals zero, because oxygen atoms cannot overlap, whereas for large r_{OO} it tends to one, i.e. the mean density is the same as in the bulk, because the ordering effect of the central oxygen has stopped acting. The calculated function is shown in Figure 6.3 where it is compared to the experimentally measured $g(r_{OO})$. The large peak near $r_{OO} = 3$ Å is due to hydrogen-bonding; the integral of this peak is related to the mean 'coordination number' of water, the number of other molecules to which it donates or from which it accepts a hydrogen bond. As can be seen, the present simulation reproduces the height, width, and position of the peak very well, indicating that it gives a good description of the dynamic nature of hydrogen bonding. The additional 'wiggles' in the experimental $g(r_{OO})$ at larger r_{OO} are due to next-nearest-neighbour effects in the

(continued...)

hydrogen-bond network. These secondary features are less well modelled, indicating that the TIP3P model for water is far from perfect. The radial distribution function was computed using structures corresponding to each 100th time-step (every 0.1 ps). The first 10 ps of the simulation was assumed to correspond to the equilibration phase and was not used—only the final 40 ps was used as the 'production' part of the simulation for generating $g(r_{OO})$. Note that the computed $g(r_{OO})$ function has some even smaller 'wiggles', e.g. near $r_{OO} = 3.5$ Å. This lack of smoothness of the function is due to numerical noise: a longer simulation would yield a smoother function.

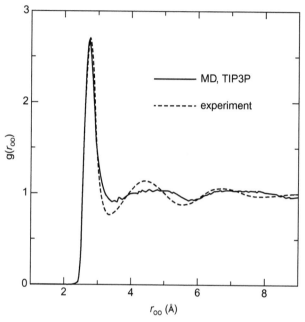

Figure 6.3 Experimental and computed radial distribution function $g(r_{OO})$ for water for $T = 298$ K. The computed function is derived from a 40 ps molecular dynamics simulation with the TIP3P forcefield.

6.4 Monte Carlo simulations

Another method to explore the potential energy surface in order to model the behaviour of the system at a given temperature is to use so-called *Monte Carlo simulations*, which derive their name from the fact that the atoms' motions are chosen in a random way—like in card games, roulette, or other casino games. The basic procedure followed is otherwise similar to that in MD (Figure 6.2). The random update procedure, called the Metropolis algorithm, is defined in such a way that it ensures that the desired statistical mechanical Boltzmann distribution is obtained over many steps. The Metropolis algorithm is schematized in Figure 6.4. As in MD, one starts from a reasonable starting structure (R_0 in Figure 6.4), and calculates the potential energy $V(R_0)$ for that structure. One then generates new structures by considering

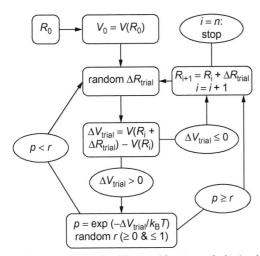

Figure 6.4 Schematic depiction of the algorithm used for Monte Carlo simulations.

random trial changes to the structure, ΔR_{trial}. If a given trial change leads to a decrease in the potential energy, the trial change is 'accepted' and a new trial change is chosen. If the trial change leads to an *increase* in the potential energy, the probability of this change for the temperature T is computed as the Boltzmann factor $p = \exp(-\Delta V/k_B T)$. The factor p is compared to a random number r chosen between 0 and 1; if $p \geq r$, the change is accepted. If not, it is rejected. This loop continues until the target number of structures n is reached.

For a given T, the set of structures R visited after a long simulation should resemble that obtained in a long MD simulation under *NVT* conditions. If one includes in the 'trial moves' changes in the volume of the periodic simulation box, one can model *NPT* conditions. One advantage of Monte Carlo methods over MD simulations is that it is also possible to include trial moves such as the addition or removal of an extra molecule to the system—this can lead to simulation of the grand canonical ensemble. The efficiency of a Monte Carlo simulation depends a lot on the way in which the random trial changes to the structure ΔR_{trial} are chosen. If these are too 'small', the likelihood that they will be accepted is high, but the range of structures explored takes a large number of steps to converge. If instead the trial changes are too large, then most of them will be rejected, and likewise the simulation will take a large number of trial steps before converging.

Monte Carlo simulations have advantages and disadvantages in comparison to MD, with some types of problem more suited to Monte Carlo and some to MD. Briefly, Monte Carlo can be preferable because it does not require computation of the gradient (which for some problems may not be readily feasible) and because the random way in which moves are generated can occasionally lead to 'large' structural changes that only occur after many steps (or never) in MD. As mentioned above, it is reasonably straightforward to model grand canonical ensembles, which is more challenging within MD. On the other hand, sequences of structures in a Monte Carlo simulation do not have a meaningful time variable, so time-dependent phenomena cannot be readily modelled.

6.5 Biomolecular simulation

In the previous sections, the two principal methods used for simulations have been described, and extensive detail has been given for one prototypical type of application, the simulation of liquid water. Although simulations of liquids remain an important tool in understanding the liquid state, by far the most common use of molecular dynamics and Monte Carlo simulations nowadays is for modelling more complex systems, and especially biomolecules such as proteins and nucleic acids. Using contemporary algorithms and computers together with molecular mechanics methods for computing the potential energy, it is possible to perform simulations with long timescales for large biomolecules or assemblies of biomolecules surrounded by solvent. At the time of writing this book, simulating systems with several tens of thousands of atoms (enough to describe a typical protein and the surrounding water) for many millions of time-steps or Monte Carlo moves was completely routine. The state of the art is for simulation of many millions of atoms and for totals of the order of 10^{12} time-steps or moves—enough to simulate milliseconds of 'real' time considering the typical time-step length of $1\text{--}2 \times 10^{-15}$ s (1 or 2 fs).

An example of a relatively small system that can readily be modelled at the time of writing using just a standard laptop computer is shown in Figure 6.5. This is the small protein crambin, which contains just 46 amino acid residues, solvated in a periodic

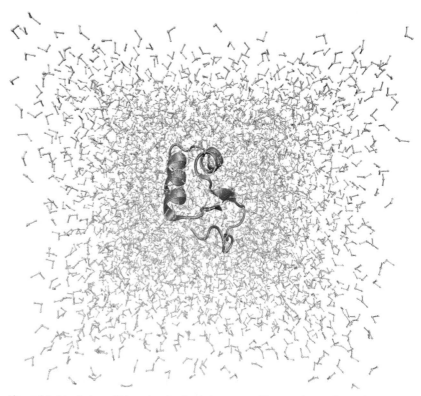

Figure 6.5 A typical small biomolecular simulation system. The protein crambin is shown in cartoon form, with recognizable α-helices, and the water molecules of the solvent are shown using balls and sticks.

box containing over 4,000 water molecules to yield a total system size of roughly 15,000 atoms, very much at the lower end of what is possible using contemporary computer resources.

Biomolecular simulation is such a large area of modern computational science that it goes well beyond the scope of this book to provide even a basic description of the main techniques used, and only a few fundamental aspects will be mentioned here.

First, simulation and related techniques play a key role in *structure determination*. X-ray crystallography plays the dominant role in determining the structures of biomolecules, together with other experimental methods such as nuclear magnetic resonance and cryo-electron microscopy. However these techniques typically do not allow the positions of all atoms to be resolved with high accuracy. Also, the models generated from experiment sometimes involve uncertainty concerning the orientation of parts of the biomolecule, to the extent that the positions of atoms residing in regions of a biomolecule that are rather flexible may not be determined at all. Modelling methods play an important ancillary role in helping to resolve these issues.

A second important area of application of simulation is in quantifying and predicting the structural variability of biomolecules. By and large, experimental techniques provide *static* structures for biomolecules, while it is known that these molecules undergo large changes in structure, which often play an important role in their function. Molecular dynamics and Monte Carlo methods intrinsically generate collections of many structures, and this can be used to provide insight into the flexibility of assemblies of biological molecules. Calculations also help to understand *solvation*—how solvent molecules arrange themselves around the different parts of a biomolecule.

A third strength of simulation methods is their ability to predict the structures adopted when two or more molecules interact with one another. A typical example of this is the case where a substrate of an enzyme, or an inhibitor, or a drug molecule, interacts with the host protein. Knowing the position where a drug molecule binds to a given protein target often plays an important role in further drug development. Even where experiment provides the overall structure of the protein, the details of the way in which smaller molecules interact with it may not be known, and simulations can assist with this. Simple molecular mechanics combined with geometry optimization can already provide some predictions concerning favoured binding positions (in a technique referred to in this context as 'docking'). Using the techniques described in Chapter 7, thermodynamic properties associated with these interactions may also be calculated.

6.6 Further reading

MD and Monte Carlo simulations are often used in conjunction with free energy methods as described in the following chapter, so the further reading material there is relevant here also.

- *The Art of Molecular Dynamics Simulation*, 2nd edition, Dennis C. Rapaport, Cambridge University Press, Cambridge, 2004. This provides a thorough description of the MD method and its applications.

- *Molecular Reaction Dynamics*, Raphael Levine, Cambridge University Press, Cambridge, 2005. This book is about experimental as well as computational aspects of reaction dynamics, but gives more insight into the use of trajectory calculations for understanding chemical reactions.

- 'Water Structure from Scattering Experiments and Simulation', Teresa Head-Gordon and Greg Hura, *Chemical Reviews*, 2002, **102**, 2651–2670. This review provides an entry into the huge literature concerning the structure of liquid water, with discussion of the role simulation plays in gaining insight into this important topic.

6.7 Exercises

6.1   By the standards of computational chemistry, the mathematics involved in simple molecular dynamics is fairly simple, so programming your own code is a feasible small project. Using only the simple Lennard-Jones potential of equation (4.5), write your own code for simulation of liquid argon based on the Velocity Verlet algorithm of equation (6.5).

6.2   Using an existing molecular dynamics code, carry out a simulation of liquid water similar to that described in the text. Use a graphical interface to visualize the results of the dynamics. In some codes, there may be options allowing you to compute the radial distribution function.

6.3   For the same system, carry out a Monte Carlo simulation, and check that it yields the same radial distribution function.

6.8 Summary

- Especially for complex systems, 'static' inspection of features of the potential energy surface is not sufficient to predict the behaviour of the system described by it.

- Simulation techniques sample the range of structures that are visited under particular experimental conditions.

- *Molecular dynamics simulations* perform this sampling by applying Newton's laws of motion to the atoms.

- *Monte Carlo simulations* instead use random changes of structure, with an algorithm for accepting or rejecting given changes that results in an overall Boltzmann distribution.

- Sampled structures can be analysed in terms of average properties, for example using *radial distribution functions*.

- The most common application of dynamical simulation methods is the study of biomolecules, their structural variability, and their interactions with other molecules.

7 Rate Constants and Equilibria

7.1 Introduction

One of the most important aims of computational chemistry is to characterize *chemical reactions*, both in terms of their thermodynamics and their kinetics. Both properties are related to relative values of the potential energy $V(R)$, using the methods of chemical *statistical mechanics*, also referred to as *statistical thermodynamics*. In this chapter, these methods to study the thermodynamics and kinetics of chemical reactions based on quantum chemistry, molecular mechanics and statistical mechanics will be discussed.

7.2 Statistical thermodynamics and equilibrium

Considering first thermodynamics, the most common quantity that you might wish to calculate is the *standard free energy change* $\Delta_r G^\ominus$ (or, as we will write it here, $\Delta_r G^0$) associated with the reaction. The standard free energy change is related to the equilibrium constant K for the reaction, $\Delta_r G^0 = -RT \ln K$ (where R is the gas constant and T the temperature). We will start by considering how the standard free energy change can be calculated for reactions of molecules in the gas phase.

The fundamental concept in statistical thermodynamics is that of *entropy*, and the second law of thermodynamics: systems evolve spontaneously in such a way that the entropy of the system and its environment is as large as possible. The entropy S is in turn connected to the disorder of the system (or of the environment), which can be measured by the number W of different configurations that are occupied in a given state of the system. The link between S and W is given by the famous Boltzmann expression, equation (7.1):

$$S = k_B \ln W \tag{7.1}$$

In many systems, W is not readily calculated, and instead the focus is on the *partition function*, q, of the system. Equation (7.2) gives the general expression for the partition function for a molecular system with a set of allowed (quantum mechanical) energy levels ε_i, each with a given *degeneracy* g_i:

$$q = \sum_i g_i \exp\left(\frac{-\varepsilon_i}{k_B T}\right) \tag{7.2}$$

The probability that a given energy level with energy ε_i is populated is given by the Boltzmann distribution of equation (7.3):

$$p_i = \frac{1}{q} g_i \exp\left(\frac{-\varepsilon_i}{k_B T}\right) \tag{7.3}$$

The importance of the partition function is that it makes it possible to evaluate the thermodynamic properties of the system. Here we will focus only on the expression for the free energy of an ideal gas formed by molecules which can each adopt a set of energy levels ε_i with degeneracies g_i. The free energy for one mole of the gas depends on p and T, and can be calculated from the molecular partition function q as shown in equation (7.4):

$$G(p,T) = G(0\ \text{K}) - RT \ln\left(\frac{q(p,T)}{N_A}\right) \tag{7.4}$$

In this expression, N_A is Avogadro's Number, and $G(0\ \text{K})$ is the absolute free energy of the species as an ideal gas at 0 K–which is simply the electronic energy for one mole of the separated molecules, corrected by their zero-point energy. Equation (7.4) provides a means to calculate $\Delta_r G^0$ for a given chemical reaction: first, you evaluate $\Delta_r G^0(0\ \text{K})$ by calculating the change in electronic energy using quantum chemistry, and the change in zero-point energy based on the vibrational frequencies, then you compute the second term based on the partition functions for all the species at the required standard p and T.

In order to evaluate the partition functions in a practical way, it is necessary to introduce some approximations, which are generally rather accurate for small or medium-sized gas-phase molecules. The first approximation is that all the allowed energy levels ε_i for the target molecules can be written as *sums* of allowed translational, rotational, vibrational, and electronic energies, equation (7.5):

$$\varepsilon_i \approx \varepsilon_{i,\text{transl}} + \varepsilon_{i,\text{rot}} + \varepsilon_{i,\text{vib}} + \varepsilon_{i,\text{elec}} \tag{7.5}$$

In this approximation, the set of possible values of $\varepsilon_{i,\text{rot}}$, the rotational energy, is taken to be the same, whatever the particular vibrational energy state of the molecule. This is an approximation, because in reality the energy in each of these types of motion depends on the other types. A rotating molecule (e.g. in a state with rotational quantum number $J = 2$) experiences centrifugal forces which cause it to distort. This means that exciting it to a higher vibrational state (e.g. moving it from $v = 0$ to $v = 1$) requires a different amount of energy than would be needed to excite a non-rotating molecule, with $J = 0$. This is well known from spectroscopic studies, and means that rotational and vibrational energy levels are coupled. Electronic energy levels are also coupled to rotational and vibrational levels. However, the coupling is in each case relatively small, so that the approximation of equation (7.5) is acceptable. It also leads to a significant simplification of equation (7.2): if all the energies are sums of separate terms, then the partition function is a *product* of translational, rotational, vibrational, and electronic parts, equation (7.6):

$$q \approx q_{\text{transl}} \times q_{\text{rot}} \times q_{\text{vib}} \times q_{\text{elec}} \tag{7.6}$$

From spectroscopy, simple and reasonably accurate expressions are known for the allowed rotational and vibrational energy levels of molecules. Closed-form expressions for the corresponding partition functions can also be obtained. For a linear molecule, with a single rotational constant B (expressed here as a wavenumber, with units cm^{-1}—if instead it is expressed as a frequency, in s^{-1}, the factor c, the speed of light, should be omitted), the rotational part q_{rot}, or 'rotational partition function', is given by equation (7.7):

$$q_{rot} = \frac{k_B T}{\sigma hcB} \tag{7.7}$$

In this expression, h is Planck's constant, and σ is a rotational symmetry number, counting the number of ways in which the molecule by rotation—so it is typically equal to 2 for symmetric linear molecules like N_2 or CO_2, and 1 for non-symmetric molecules such as HCl. The rotational constant B is a simple function of the atomic masses and of the bond length, so it can be predicted provided that you know the structure of the molecule. For example, for a diatomic molecule X–Y with bond length R_{XY} and atomic masses m_X and m_Y, the rotational constant B_{XY} is given by (equation (7.8)):

$$B_{XY} = \frac{h}{8\pi^2 cI_{XY}} \tag{7.8}$$

Where c is the speed of light, and I_{XY} is the moment of inertia of the molecule, equation (7.9):

$$I_{XY} = \frac{m_X m_Y}{m_X + m_Y} R_{XY}^2 \tag{7.9}$$

For non-linear molecules, with three rotational constants A, B, and C, equation (7.7) should be replaced by equation (7.10):

$$q_{rot} = \frac{1}{\sigma} \left(\frac{k_B T}{hc} \right)^{\frac{3}{2}} \left(\frac{\pi}{ABC} \right)^{\frac{1}{2}} \tag{7.10}$$

For the vibrational part q_{vib} in equation (7.6), we consider first diatomic molecules, in which there is only one possible vibration, with frequency v. Using the harmonic approximation already discussed in Chapter 5, whereby stretching or compressing the bond leads to an increase in energy that is proportional to the square of the magnitude of the distortion, it is possible to derive the simple expression of equation (7.11). Especially for the low energy levels that contribute most to q_{vib} at low and moderate T, the harmonic approximation is accurate.

$$q_{vib} = \frac{1}{1 - \exp \frac{-hv}{k_B T}} \tag{7.11}$$

For polyatomic molecules, with multiple frequencies v_i obtained by diagonalizing the Hessian matrix, and still using the harmonic approximation, this can be generalized to equation (7.12):

$$q_{vib} = \prod_i \frac{1}{1 - \exp \frac{-hv_i}{k_B T}} \tag{7.12}$$

For the *translational* part q_{trans} it is also possible to obtain an analytical expression, and this is most conveniently done by describing the gas molecules as particles in a box. The allowed energy levels depend on the dimensions of the box, in all three directions, in other words on the volume V of the box—and so does the partition function, which can be expressed as equation (7.13):

$$q_{trans} = \left(\frac{2\pi m k_B T}{h^2} \right)^{\frac{3}{2}} V \tag{7.13}$$

Where m is the mass of the molecule, and V is the volume of the box that it is moving in, which is equal to RT/p^0 when considering a mole of ideal gas at the standard pressure p^0.

For many molecules, only one electronic energy level is populated at even quite high temperatures, which means that q_{elec} is usually equal to one. For molecules with unpaired electrons, q_{elec} is however equal to the spin multiplicity (e.g. 2 for a radical or spin doublet, or 3 for a spin triplet such as O_2).

Putting all of this together yields a procedure to calculate the overall partition function for a gas-phase molecule, and thereby the corresponding free energy. In turn, this provides a route to calculate $\Delta_r G^0$ and thereby K. All the molecular properties needed—the rotational constants A, B, or C; the vibrational frequencies v_i; and the molecular mass m are either known or can be computed from the results of a quantum chemical (or molecular mechanics) geometry optimization and vibrational frequency calculations. Evaluating the free energy using equations (7.4), (7.6), and (7.7) or (7.10), (7.12), and (7.13) is not particularly computationally challenging and does not strictly speaking require a computer. This simple calculation is often performed automatically by quantum chemical codes whenever vibrational frequencies are computed.

Worked example: the $N_2 + 3\,H_2 \rightleftharpoons 2\,NH_3$ equilibrium

The synthesis of ammonia from the elements, the Haber–Bosch process, is one of the most important transformations carried out by the chemical industry. To illustrate the methods discussed above, we will calculate K_p, the equilibrium constant for the reaction, at a temperature typical for the synthesis of ammonia from the elements. Equation (7.14) shows how the standard free energy change relates to the equilibrium constant K_p, which itself relates to the equilibrium partial pressures for the reactants and product. Equation (7.14) also includes the standard state pressure p^0, included so that the overall equilibrium constant is dimensionless. Even though K_p is dimensionless, it implicitly refers to a certain choice of units for the pressures—here, the choice will be to use $p^0 = 1$ bar as the reference.

$$\Delta_r G^0 = -RT \ln K_p = -RT \ln \left(\frac{p_{NH_3}^2 (p^0)^2}{p_{N_2} p_{H_2}^3} \right) \tag{7.14}$$

(Note that it is possible to define K_p as the square root of the above value, by halving the stoichiometric constants in the expression for the reaction. This leads to a value of $\Delta_r G^0$ that is half that obtained here.) The first ingredient needed to calculate K_p for this reaction is the free energy change for the reaction between the gases, considered at 0 K. This is simply the electronic total energy difference between products and reactants, corrected by the zero-point energy difference.

(continued...)

For this simple reaction, the electronic structure calculations are straightforward to carry out. Table 7.1 shows some calculated values for N_2, H_2, and NH_3. As well as the rather inaccurate Hartree–Fock level, results are shown for MP2 and CCSD(T), the latter using the large aug-cc-pVQZ basis set. All energies are single-point energies computed at the B3LYP/6-31G(d) structures. As B3LYP yields a fairly accurate structure, calculating the MP2 and CCSD(T) energies at this structure rather than carrying out geometry optimization is a good approximation, which has the benefit that it decreases the computational expense. Because this reaction is *isogyric* (though not *isodesmic*), the difference between the HF and CCSD(T) calculated reaction energy is not enormous, though it is still sizeable in chemical terms.

Table 7.1

	$E(H_2)$	$E(N_2)$	$E(NH_3)$	$\Delta_r E$ (/kJ mol^{-1})
HF	−1.13345	−108.98980	−56.22326	−61.2
MP2	−1.16671	−109.39363	−56.47757	−74.4
CCSD(T)	−1.17387	−109.40718	−56.49563	−77.2

Note: all total energies are in hartree and were calculated at structures optimized with the B3LYP/6-31G(d) level of theory. The indicated reaction energies include a correction for zero–point energy, of +86.8 kJ mol^{-1}, based on B3LYP/6–31G(d) vibrational frequencies.

With this in hand, you can then calculate the other required ingredient, the partition functions for reactants and products, using the calculated properties given in Table 7.2, obtained from the B3LYP/6-31G(d) structures and vibrational frequencies. For this calculation (Table 7.2), we use $T = 800$ K, a typical operating temperature for the Haber process for synthesis of ammonia from the elements. The standard state pressure of $p^0 = 1$ bar (10^5 Pa) corresponds to a molar volume at $T = 800$ K of 66.5 l or 0.0665 m^3 ($V_{mol} = RT/p^0$). The symmetry numbers for H_2, N_2, and NH_3 are respectively 2, 2, and 3. In this case, the statistical thermodynamics calculation has been done 'by hand' using the masses, rotational constants, and vibrational frequencies. In practice, this is not necessary: quantum chemical codes automatically calculate partition functions when calculating vibrational frequencies, so most of the work is done for the user.

Table 7.2

Species	H_2	N_2	NH_3
m/amu	2.016	28.006	17.027
m/kg	3.348×10^{-27}	4.651×10^{-26}	2.827×10^{-26}
q_{trans}	8.09×10^{29}	4.19×10^{31}	1.99×10^{31}
A,B,C/cm^{-1}	60.6	1.97	9.77, 9.77, 6.33
q_{rot}	4.59	141	315
ν/cm^{-1}	4451	2458	1132, 1727 (×2), 3437, 3568 (×2)
q_{vib}	1	1.01	1.27
q_{elec}	1	1	1
ln (q/N_A)	15.63	23.02	23.30
$G - G(0)$/kJ mol^{-1}	−104.0	−153.1	−155.0

The overall reaction free energy for $T = 800$ K and $p = 1$ bar is thereby predicted to be $\Delta_r G^0(800$ K$) = −77.2 + (2 \times −155.0 - (−153.1 + 3 \times −104.0)) = 77.9$ kJ mol^{-1}. This is

(continued...)

equivalent to a value of K_p of 8.2×10^{-6} (this is a dimensionless quantity, but it implicitly refers to equilibrium pressures in bar). The reaction is very unfavourable in free energy terms at this temperature and pressure. The experimental thermodynamics of nitrogen, hydrogen, and ammonia at different temperatures are well known. Retrieved from the NIST database, http://webbook.nist.gov/chemistry/, in June 2017, they yield $\Delta_r G(800\,K) = +77.2$ kJ mol^{-1}, or $K_p = 9.0 \times 10^{-6}$, very close to the calculated value. The extremely good agreement is partly fortuitous (errors that cancel each other out), and partly results from the fact that for this simple gas-phase reaction the statistical mechanics approach used is very accurate. Also, the electronic structure theory method applied is very good both in terms of the correlation treatment and the basis set. It should be remembered from Chapters 2 and 3 that achieving an error of 5 kJ mol^{-1} or less for bond energies is very challenging. An error by 5 kJ mol^{-1} is equivalent to an error for the equilibrium constant by a factor of $\exp(\Delta\Delta G/RT) = 2$ at 800 K; at room temperature the same error has a bigger impact, by a factor of 7.5. With less accurate electronic structure theory methods, errors by 20 kJ mol^{-1} are not unusual, leading to errors on predicted equilibrium constants at room temperature by a factor of over 3,000. The error due to the statistical mechanics treatment is usually less severe.

7.3 Transition state theory

Statistical thermodynamics can also be used to calculate rate constants, at least in an approximate way. The standard free energy difference between the reactants and the transition state to the reaction, written as $\Delta G^\ddagger$ and referred to as the free energy of activation, provides an estimate to the rate constant for the reaction through *transition state theory*, which in its simple form is written as the Eyring equation, equation (7.15):

$$k \cong k_{TST} = \frac{k_B T}{h} \times \frac{1}{c_{std}^{\Delta n}} \times \exp\left(\frac{-\Delta G^\ddagger}{RT}\right) \tag{7.15}$$

In this equation, k_B is Boltzmann's constant, T is temperature, h is Planck's constant, c_{std} is the standard concentration assumed when calculating the translational partition functions, Δn is the change in number of molecules between the reactants and the TS (e.g. 0 for a unimolecular reaction, or 1 for a bimolecular reaction). The term with the standard concentration is often omitted since its numerical value is 1, and it is present merely to make units consistent. The key necessary input for equation (7.15), the activation free energy $\Delta G^\ddagger$, can be calculated in the same way as a standard free energy change for a chemical reaction. When calculating the vibrational partition function q_{vib} for the TS using equation (7.12), the imaginary frequency corresponding to the reaction coordinate is omitted from the analysis, as it instead contributes the factor $k_B T/h$ to the rate constant.

Transition state theory calculation of the rate constant for the $OH + CH_4 \rightarrow H_2O + CH_3$ reaction

To illustrate the power of equation (7.15), we consider the reaction of the OH radical with methane to yield methyl radical and water. This is the main reaction leading to removal of methane from the atmosphere, and is also the main source of water molecules in the upper atmosphere, the stratosphere. Given its importance, there

(continued...)

have been many experimental studies, and the consensus rate constant at 298 K is 6.68×10^{-15} cm^3 molecule^{-1} s^{-1}. The structure of the reactants and the TS have been optimized at the MP2 level of theory using the 6-311G(d,p) basis set, and the energy of each of these species has been computed using ROHF, MP2, CCSD, and CCSD(T) together with the aug-cc-pVTZ basis set. As in the example above, the molecular masses, rotational constants, and harmonic vibrational frequencies have been used to compute the partition functions. From the partition functions, the standard free energy difference between the reactants and the TS is obtained—this has been done by the quantum chemical code in the present case. All computed values are shown in Table 7.3.

Table 7.3

	OH	CH$_4$	TS	Δ(reac-TS)
HF	−75.41677	−40.21355	−115.58257	125.4
MP2	−75.62655	−40.41443	−116.02829	33.3
CCSD	−75.63939	−40.43437	−116.05805	41.3
CCSD(T)	−75.64537	−40.44093	−116.07504	29.5
z.p.e.	0.00878	0.04552	0.05176	−6.7
G − G(0)	−0.01691	−0.01835	−0.02759	20.1
G (298 K)	−75.65335	−40.41377	−116.05087	43.1

All energies are in hartree, except for those relating to the difference between the reactants and the TS, which are in kJ mol^{-1}. The translational partition function of equation (7.13) is calculated using a volume equal to the molar volume of a gas at $p = 1$ atm ($= 1.013$ bar) and $T = 298.15$ K. Hence the free energy also refers to this reference point.

Using this calculated value of the activation free energy, application of equation (7.15) yields $k = 1.81 \times 10^5$ atm^{-1} s^{-1}. The units obtained result from the choice of standard state in equation (7.13) is that occupied by a mole of ideal gas at a pressure of 1 atm, $V = RT/p^0$. To obtain the rate constant in the units used by gas-phase kineticists, the value above needs to be multiplied by the conversion factor atm cm^3 molecule^{-1}, that is, it needs to be divided by the concentration in molecules per cm^3 of an ideal gas at $p^0 = 1$ atm. This is equal to 2.461×10^{19}, yielding $k = 7.34 \times 10^{-15}$ cm^3 molecule^{-1} s^{-1}, only 10% larger than the experimental value. As for the calculation of the equilibrium constant for the ammonia synthesis in the worked example of the previous box, this good agreement is partly fortuitous. In fact, this is even more the case here, since while equation (7.14) relating the free energy to the equilibrium constant is exact, the Eyring equation, equation (7.15), is an approximation, introducing an additional potential source of error between the calculation and experiment. For example, equation (7.15) does not account for possible quantum mechanical tunnelling in the transfer of hydrogen from carbon to oxygen in the TS. Still, the good agreement is also due to the basic accuracy of statistical thermodynamics for this system and to the good accuracy of the quantum chemical calculations. As for equilibria, relatively small errors of 5 or 20 kJ mol^{-1} for $\Delta G^{\ddagger}$ will cause an error by a factor of 7.5 or over 3,000 for k at 298 K.

7.4 Free energies from MD and MC simulations

In the previous two sections, it has been shown that partition functions derived from quantum chemically calculated properties, and based on quantum mechanical expressions for allowed rotational, translational, and vibrational energies, can lead to accurate predictions of relative free energies. The same approach based on the harmonic oscillator approximation can in principle also be applied to properties predicted with molecular mechanics. However the equations shown previously apply only to molecules existing as a single conformer. For small numbers of conformers, i.e. for relatively small molecules, it is possible to extend equation (7.2) so that the summation extends over all of these local minima on the potential energy surface. In practice, though, as the system size increases, the number of conformers does too, and at some point it becomes impossible to take them all into account, with the consequence that the approach outlined above becomes inapplicable. Hence, for calculating free energies based on the sort of large system treated with molecular mechanics, another methodology is needed.

It should be noted that even for molecular systems treated with quantum chemistry, the problem just mentioned becomes significant for larger molecules, which can lead to quite severe errors in calculated free energies. Such errors arise when calculations have not identified the lowest-energy minimum (remember that standard geometry optimization only yields local minima, not the overall or global minimum), but even where the global minimum is used, neglecting relatively low-lying other conformers can have a significant deleterious effect on calculated free energies.

For extended systems, a wholly different starting point is used to express the partition function, in place of equation (7.2). Instead of focusing on the allowed quantum energy levels of the target molecule, you consider the possible classical positions R and velocities v that can be adopted by the atoms in the system. This leads to equation (7.16) for the classical partition function of a system of N atoms:

$$q = \frac{1}{h^{3N}} \iint \exp\left(\frac{-\left\{V(R) + \sum_i \frac{m_i v_i^2}{2}\right\}}{k_B T}\right) dR dv \qquad (7.16)$$

Where $V(R)$ is the potential energy, and the integral is taken over all the possible coordinates R and velocities v_i for all the atoms in the system. The factor of $1/h^{3N}$ (where h is Planck's constant) can often be omitted since it is a constant, which usually cancels out when considering relative free energies. It is present in order to ensure that the partition function has the same numerical value as that obtained from the quantum mechanical equation (7.2) in cases where it can be evaluated in both ways.

The partition function of equation (7.16) can be rewritten as a product of one integral over the positions R and one over the velocities v; and the latter can be evaluated analytically. This second part usually has the same value for all systems with the same atomic composition. As a consequence, its contribution to *relative* free energies usually cancels out, and it can be ignored. This means that the partition function can be rewritten as equation (7.17):

$$q = C \int \exp\left(\frac{-V(R)}{k_B T}\right) dR \qquad (7.17)$$

By analogy to equation (7.3), the probability that a given position R will be adopted can also be written as in equation (7.18):

$$p(R) = \frac{1}{q} \exp\left(\frac{-V(R)}{k_B T}\right) \qquad (7.18)$$

Neither of these expressions can be evaluated exactly, because, except for small systems, there are simply far too many possible sets of coordinates R to consider, and no analytical means to evaluate the integral given the complexity of the potential energy surface. Still, they do serve as a useful framework for considering the outcome of simulations and for deriving approximate expressions for thermodynamic properties such as free energies. This can be done based on molecular dynamics or Monte Carlo simulations, with the assumption that the probability with which structures are visited during such simulations is proportional to that expressed by equation (7.18).

This approach can be illustrated using a simple example, using an MD simulation of liquid n-butane near its boiling point, at 273 K. A box of 50 butane molecules (so 700 atoms) is constructed using the same approach as described earlier for water molecules. Based on a target density of 0.6016 g cm^{-3}, the molecules are simulated using periodic boundary conditions in a cubic box with length 20.0125 Å. After suitable optimization and equilibration, the system is simulated for a total of 120 ps, using a time-step of 1 fs, and keeping copies of the structures every 100 fs (or 0.1 ps, so 1,200 structures are saved). The MM3 forcefield is used.

The simulation time used here is not even remotely long enough to sample *all* possible structures for the 700 atoms. This is because the conformational space available even for this fairly modestly sized system is unimaginably large. For example, each butane molecule can translate in all three dimensions in the box, can also rotate around three axes, and has conformational flexibility with *anti* and *gauche* conformers. Full sampling for equation (7.17) would require not only that each molecule explore all of these positions, but also that all *combinations* of the positions of all molecules be explored. This will not be anywhere close to being realized with the simulation time used here.

Nevertheless, the system can perhaps be expected to sample a subset of the available space in a representative way, so that the probabilities are proportional to those of equation (7.18). Even though the full 2,100-dimensional space of values of R will not be visited, the distribution within a lower-dimension subspace may be correctly reproduced. If this is the case, then perhaps an adapted version of equation (7.4) can be used to compute free energies. For example consider just the butane dihedral angle $\theta(C_1–C_2–C_3–C_4)$ in the molecules. Does this adopt all possible values during the simulation? Figure 7.1 shows a plot of the values of θ explored by all fifty molecules in the sample during the simulation. This distribution appears to make excellent sense with what might be expected: the most frequently observed values of θ correspond to the most stable *anti* conformer ($\theta = 180°$) and the two enantiomeric *gauche* conformers ($\theta = 60°$ and $300°$). The spread around the equilibrium values reflects the thermal motion. The plot is almost but not exactly symmetric around $\theta = 180°$, with the deviations due to the limited timescale of the simulation and the way the θ values have been grouped to produce the histogram. Looking in more detail at the simulation output, it can be seen that θ values near $120°$ and $240°$, i.e. near the TSs for converting the *gauche* conformers to the *anti* conformer (where each methyl is eclipsed with a hydrogen), are also visited reasonably frequently during the simulation. Values of $\theta \cong 0°$, near the TS for interconverting the two *gauche* forms, are visited less often, which makes sense since this TS lies higher in energy, as it involves an eclipsed interaction between the two methyl groups.

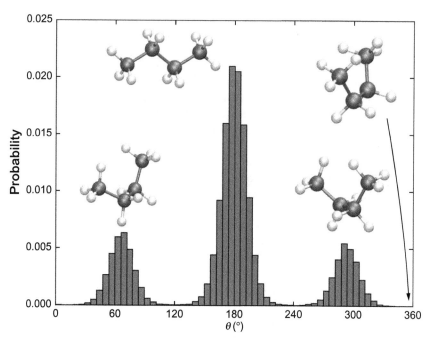

Figure 7.1 Distribution of dihedral angles $\theta(C_1-C_2-C_3-C_4)$ observed during a molecular dynamics simulation of a box of n–butane at T = 273 K, using the MM3 forcefield. The structures of the two enantiomeric *gauche* structures, the *anti* structure, and the TS linking the two *gauche* structures are superimposed.

The distribution of dihedral angles observed during the simulation is reasonably well converged, even though all the structures that correspond to the overall partition function were not (and reasonably speaking, could not have been) visited. Considering all 50 molecules and all 1,200 saved structures, the molecules can be classified as being *gauche, g,* or *anti, a,* based on the value of θ. The *gauche* structures can be further classified as belonging to one of the two enantiomeric *gauche* structures g^+ and g^-. For example, a structure can be considered to be g^+ for θ between 0° and 120°, *a* for θ between 120° and 240° or g⁻ for θ between 240° and 360°. Analysis of the simulation shows that for the 50 molecules, each observed 1,200 times, the total of 50 × 1,200 = 60,000 structures break down as $n(g^+) = 11,599$, $n(g^-) = 9,764$ and $n(a) = 38,637$. These values can be taken to be proportional to the partition functions for these different conformers. With this approximation, and using equation (7.4), this makes it possible to estimate the relative free energies of the three conformers (the anti conformer is taken as reference so $\Delta G(a) = 0$), as shown in equations (7.19) and (7.20):

$$\Delta G(g^+) = -RT \ln\frac{n(g^+)}{n(a)} = 2.73 \text{ kJ mol}^{-1} \tag{7.19}$$

$$\Delta G(g^-) = -RT \ln\frac{n(g^-)}{n(a)} = 3.12 \text{ kJ mol}^{-1} \tag{7.20}$$

The small difference between the two values—which should be identical, given the symmetry of the problem—reflects the numerical error associated with the finite length of the simulation. Another manifestation of this error can be seen by using only the first 60 ps of the simulation to recompute equations (7.19) and (7.20). The relative free energies obtained then have values of respectively 3.09 and 2.59 kJ^{-1}. Taking just the first 50 structures (so 5 ps of the simulation) yields values of 3.34 and 2.40 kJ mol^{-1}. Still all of these numbers are less than 1 kJ mol^{-1} different from the 120 ps values, suggesting that the numerical error is acceptable, and the simulation gives a reasonably good approximation to the relative values of the partition functions for the different conformers.

An alternative way of analysing the spread of θ values is to plot the function of equation (7.21):

$$\mu(\theta) = -RT \ln n(\theta) + C \tag{7.21}$$

Where $n(\theta)$ is the number of times that the dihedral angle is close to a given value θ, and C is a constant. Like equations (7.19) and (7.20), equation (7.21) relies on the link between the partition function and the free energy also displayed in equation (7.4). As in equations (7.19) and (7.20), it is assumed that $n(\theta)$ is proportional to the full partition function of equation (7.17) evaluated for those structures R characterized by the given value of θ. The quantity $\mu(\theta)$ is referred to as a *potential of mean force* or *pmf*, and can be plotted as a function of θ, yielding Figure 7.2 below, where C has been chosen such that $\mu = 0$ kJ mol^{-1} for $\theta = 180°$. Note that values of θ below 0° and above

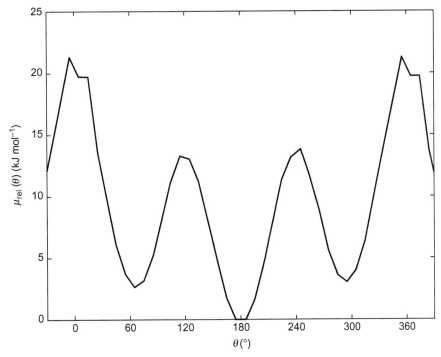

Figure 7.2 Potential of mean force $\mu(\theta)$ associated with the distribution of dihedral angles θ (C_1-C_2-C_3-C_4) from Figure 7.1.

360° have been added to make it easier to read the graph (using $\mu(\theta) = \mu(\theta+360°)$). The name 'potential of mean force' arises because the same quantity can also be obtained by averaging over the derivative of the potential energy with respect to θ (which is a force). The quantity μ is related to a free energy, so from Figure 7.2 it can be seen that the *gauche* conformers lie roughly 2.5 kJ mol^{-1} above the *anti* conformer, with the $g \to a$ TSs at about 12 kJ mol^{-1}, and the $g^+ \to g^-$ TS near 20 kJ mol^{-1}. Correct extraction of activation free energies from pmfs requires a more complicated analysis, because here we have used $n(\theta)$ for individual values of θ, whereas in fact, as shown in the preceding paragraphs, the free energy depends on sums of $n(\theta)$ across whole basins. The detailed differences between pmfs and free energies go beyond the scope of this book.

It is interesting to compare the results obtained here with those that are obtained with the approach used earlier in the chapter for computing G. This cannot be directly applied to the present problem for the reasons described at the beginning of the section, but it can be applied to a butane molecule in vacuum, using the same MM3 potential energy surface. The relative free energies calculated in this way for the anti minimum, gauche minimum, TS for anti-gauche conversion ($\theta \cong 120°$ or $240°$), and TS for gauche–gauche conversion ($\theta \cong 0°$) are 0.0, 3.9, 10.1, and 22.2 kJ mol^{-1}. These values are similar to those obtained from the pmf, but not identical. The reasons for the differences include the fact that the 'quantum' approach for calculating partition functions includes aspects, such as zero-point energy, that are omitted in the MD simulation approach, and the mentioned fact that pmfs do not exactly map onto free energies. Also, of course, one set of values was obtained for liquid butane, and the other for the molecule in the gas phase. This is, however, a fairly small difference for the present case.

Looking more closely, Figures 7.1 and 7.2 illustrate one of the major problems in trying to extract free energies from simulations. This is that configurations corresponding to high potential energies are seldom visited during the simulation: these are 'rare events'. In the present case, only one case of a dihedral angle between 350° and 360° was recorded during the whole simulation—and the range 0° to 10° was visited only twice. This is roughly in line with what would be expected from transition state theory: for $\Delta G^* = 20$ kJ mol^{-1} at 273 K, equation (7.15) predicts $k = 8 \times 10^8$ s^{-1}. Considering that there were 50 molecules in the sample, and that the simulation was 120 ps = 1.2 $\times 10^{-10}$ s long, you might have expected to see roughly $50 \times 8 \times 10^8 \times 1.2 \times 10^{-10} = 5$ crossings of the $g^+ \to g^-$ TS. If crossing the TS region (θ between $-10°$ and $10°$) took about 10 fs (a typical value), then by sampling structures every 100 fs, only one such barrier crossing would be recorded, as was indeed the case. The very low number accounts for the 'spiky' nature of the pmf near $\theta = 0°$ and 360°.

Considering only the extent to which the different minima were visited, the simulation can however be considered to have been close to convergence: there have been enough barrier crossing events so as to ensure that all relevant low-free energy regions have been visited, and to reach a state of near-equilibrium in terms of the relative populations of these regions. In fact, even the key TS regions have been sampled (albeit only just in the case of the $g^+ \to g^-$ TS).

This is a very favourable outcome, which partly reflects the details of the particular case. For example, one key TS, that connecting the g^+ and g^- conformers, does not need to be crossed at all to interconvert conformers. The lower-energy $g^+ \to a$ and $a \to g^-$ TSs provide an alternative route. Also, when simulating a liquid with a model containing fifty molecules but focusing on an internal property of each molecule, we effectively carry out fifty simulations for the cost of one.

In most problems of interest, though, the situation is much less favourable. Typically, when simulating large molecules such as proteins, only one target molecule is present in the simulation system, together with solvent. And barriers of many tens of kJ mol^{-1} exist between conformers. At $T = 298$ K, transition state theory suggests that a barrier of 20 kJ mol^{-1} would be crossed on average once every 500 ps, dropping to once every 30 ns for $\Delta G^\ddagger = 30$ kJ mol^{-1} and once every 1.7 μs for $\Delta G^\ddagger = 40$ kJ mol^{-1}. In such cases, if you start from a structure that is on the 'wrong' side of the barrier, the simulation might never reach the correct region. This problem occurs for MD simulations, but also to some extent for Monte Carlo simulations, since the 'jump' in structure needed to cross a barrier is also very unlikely to be selected randomly.

7.5 Enhanced sampling techniques

In order to deal with the problems associated with 'rare events' discussed above, it is necessary to increase the extent to which sampling of the potential energy surface is carried out. One obvious way to do this is to lengthen the simulation time. More efficient MD and Monte Carlo algorithms and faster computers mean that simulations of large systems with μs and even ms timescales have become possible. In some cases, long total simulation times can be obtained by running multiple shorter simulations (though if they all start in the incorrect region of the potential energy surface, and there is a large barrier separating it from the 'correct' region, this still will not guarantee correct sampling).

Increased sampling can also be achieved in a large number of other ways. One option, known as *replica exchange* methods, is to perform part of the simulation using one or more different (and higher) temperatures. As stated above, at 298 K, crossing a 40 kJ mol^{-1} barrier might occur on average once every 1.7 μs—whereas at 373 K, it would be on average once every 50 ns. Sophisticated methods can be used to switch temperature repeatedly, with transitions between different low-energy regions being enhanced at the higher temperatures, but populations of different states being evaluated at lower temperatures.

Another option is to modify (or *bias*) the potential energy surface, so as to lower the energy barriers. The so-called biasing potential can be specified in advance of the simulation, or it can be modified during the simulation so as to respond to the current state of the system. Also, you can attempt to generate one biasing potential that allows efficient and converged sampling of the whole region of interest, or you can run multiple simulations with different biasing potentials. Of course, the distribution over structures observed in the biased simulations cannot be directly used to compute relative free energies or pmfs. Instead, the effect of the biasing potential must be removed. The details of how this is done depend on how the bias was introduced and go beyond the scope of this book. Important methods such as *metadynamics* and *umbrella sampling* are based on this general idea of using a biased potential.

Another family of techniques frequently used to compute relative free energies from MD or MC simulations falls under the general name of 'free energy perturbation'. Consider a system that can exist in two different states A and B, described by the potential energy functions V_A and V_B. Their partition functions q_A and q_B can each be written as in equation (7.17), and the free energy difference between the two states is given by $\Delta G = -RT \ln (q_B/q_A)$. Assuming that both systems depend on the same set of coordinates R, and that the masses of the atoms are the same, so that the prefactor

C of equation (7.17) is the same for both systems, the ratio of partition functions can also be written as equation (7.22):

$$\frac{q_B}{q_A} = \frac{\int \exp\left(\frac{-\{V_B(R) - V_A(R)\}}{k_B T}\right) \times \exp\left(\frac{-V_A(R)}{k_B T}\right) dR}{\int \exp\left(\frac{-V_A(R)}{k_B T}\right) dR} \tag{7.22}$$

The final term in the integrand of the numerator and the denominator together form the probability of adopting the coordinates R in state A, as defined in equation (7.18), so equation (7.22) can be rewritten in the following form, equation (7.23):

$$\frac{q_B}{q_A} = \int \exp\left(\frac{-\{V_B(R) - V_A(R)\}}{k_B T}\right) \times p_A(R) dR \tag{7.23}$$

This equation can be read in the following way: the ratio of partition functions is given by the integral of the Boltzmann factor of the potential energy difference $(V_B - V_A)$ at a set of structures sampled for state A. Once again, it is clearly not possible to sample *all* possible structures and thereby evaluate this ratio of partition functions exactly. Still, under some conditions, it is possible to obtain a good approximation. An MD or Monte Carlo simulation can be used to sample a set of structures for state A approximately corresponding to the Boltzmann distribution. In favourable cases, the factor $\exp(-\Delta V/k_B T)$ in equation (7.23) does not vary too much, so that the integral is roughly converged for a relatively short simulation. Equation (7.23) can then be used to obtain the free energy perturbation expression for the difference in free energy, equation (7.24):

$$\Delta G_{A \to B} = -RT \ln \left\langle \exp\left(\frac{-\{V_B(R) - V_A(R)\}}{k_B T}\right) \right\rangle_A \tag{7.24}$$

In this equation, the angular brackets $< \ldots >$ with the subscript A denote averaging over a suitably large number of structures R during a simulation of state A. The expression is symmetric with respect to states A and B, so that the free energy change could also be computed based on a simulation of state B.

In many cases of interest, $\Delta V = V_B - V_A$ depends quite strongly on the coordinates R, with the favourable structures for A, that are extensively visited during the simulation, having low V_A but relatively high V_B so ΔV is large and yields a small value of the exponential term $\exp(-\Delta V/k_B T)$. Conversely, the most favourable structures for B, for which ΔV is smaller and the exponential term much larger, may be visited only very seldom. Because of the steep variation of the exponential function, the few structures with smallest ΔV can make a disproportionate contribution to ΔG and obtaining a reliable estimate of the latter may require a very long simulation, or may even not be possible at all.

When this happens, the transformation from state A to state B can be broken down into a number of steps, and equation (7.24) can be applied to each of them. This can be written conveniently in terms of a parameter λ that is equal to 0 for state A and 1 for state B. Instead of applying equation (7.24) to $\Delta G_{A \to B} = \Delta G_{\lambda = 0 \to \lambda = 1}$, you apply it first to a smaller change, such as that between the states with $\lambda = 0$ and $\lambda = 0.2$. Then it can be applied to the change $\lambda = 0.2 \rightarrow \lambda = 0.4$, and so on. The overall free energy change is then given by equation (7.25):

$$\Delta G_{A \rightarrow B} = \sum_{\lambda_i = 0}^{\lambda_{i+1} = 1} \Delta G_{\lambda_i \rightarrow \lambda_{i+1}} \tag{7.25}$$

In this expression, the free energy change for each step is computed based on structures sampled using the potential energy surface for either the starting or the end point of the step. By choosing enough intermediate values of λ, the change in the potential energy expression between each of the adjacent pairs of λ values can be small enough that equation (7.24) converges rapidly. As a test, the whole procedure can be run 'backwards' from B to A—if the conditions for convergence are met, the overall free energy change in both directions should be consistent.

Free energy perturbation: Example

This procedure is best understood by taking an example, which here will be the classic case of the difference in the free energies of solvation in water of ethane (C) and methanol (D). System A corresponds to an ethane molecule dissolved in water as well as an isolated methanol molecule, and system B corresponds to a methanol molecule dissolved in water as well as an isolated ethane molecule. Each solute is treated according to the OPLS-AA forcefield, using periodic boundary conditions in a cubic box of length 24.662 Å filled with 496 water molecules, described using the TIP3P potential, and one molecule of the solute. The structure of the solute (or the intermediate state with $\lambda \neq 0$ or 1) is fixed during each of the simulations used to evaluate the free energy change.

For intermediate states where λ is neither 0 nor 1, the partial charges and Lennard-Jones parameters σ and ε for the solute atoms, as well as the coordinates for the solute atoms, are assigned to lie between the values appropriate for ethane and those appropriate for methanol. For example, in the OPLS-AA forcefield, the C and H atoms of ethane have charges of −0.18 and +0.06, respectively. In methanol, the charges on the C, O, H (in CH_3), and H (in OH) atoms are, respectively, +0.145, −0.683, +0.04, and +0.418. Also, two of the H atoms in ethane are not present in methanol, but in the simulation, they are treated as 'atom' sites with charges of 0. For the intermediate state with $\lambda = 0.5$, the forcefield uses for example a charge of $−0.18 + 0.5 \times (0.145 − (−0.18)) = −0.0175$ for the C atom, of $−0.18 + 0.5 \times (−0.683 − (−0.18)) = −0.4315$ for the carbon atom that is being 'converted' to oxygen, and $0.06 + 0.5 \times (0. − 0.06) = 0.03$ for the H atom that is 'vanishing'. A similar scaling is done for the σ and ε values, and for the coordinates (the ethane and methanol molecule are oriented in the same way, so that each atom 'moves' as little as possible from $\lambda = 0$ to $\lambda = 1$).

Several simulations each of 30 ps duration are performed, each involving a different change in λ. The first 2 ps are considered to allow for equilibration of the system, with the remaining 28 ps used for computation of free energy differences. In this case, ten equal increments of λ from 0 to 1 were used. In each simulation, the initial value λ was used to define the forcefield for the simulation, and after every few dynamics steps, the energy for the current structure is computed using the forcefield for the next value of λ. This yields one value for the difference in potential energy ΔV, which after extensive sampling is averaged using equation (7.24) to yield $\Delta G(\lambda_i \rightarrow \lambda_{i+1})$. The same procedure is also run in reverse from $\lambda = 1$ to 0, and the results are shown in Table 7.4.

(continued...)

Table 7.4. Calculated free energy increments (in kcal mol^{-1}) for stepwise free energy perturbation treatment of the difference in solvation free energy between ethane ($\lambda = 0$) and methanol ($\lambda = 1$). Results for the reverse process are also shown.

Step	$\Delta G(\lambda_i \rightarrow \lambda_{i+1})$	Step	$\Delta G(\lambda_i \leftarrow \lambda_{i+1})$
$\lambda = 0 \rightarrow \lambda = 0.10$	−0.240	$\lambda = 0 \leftarrow \lambda = 0.10$	0.296
$\lambda = 0.10 \rightarrow \lambda = 0.20$	−0.290	$\lambda = 0.10 \leftarrow \lambda = 0.20$	0.149
$\lambda = 0.20 \rightarrow \lambda = 0.30$	−0.159	$\lambda = 0.20 \leftarrow \lambda = 0.30$	0.239
$\lambda = 0.30 \rightarrow \lambda = 0.40$	−0.317	$\lambda = 0.30 \leftarrow \lambda = 0.40$	0.267
$\lambda = 0.40 \rightarrow \lambda = 0.50$	−0.339	$\lambda = 0.40 \leftarrow \lambda = 0.50$	0.459
$\lambda = 0.50 \rightarrow \lambda = 0.60$	−0.578	$\lambda = 0.50 \leftarrow \lambda = 0.60$	0.525
$\lambda = 0.60 \rightarrow \lambda = 0.70$	−0.703	$\lambda = 0.60 \leftarrow \lambda = 0.70$	0.762
$\lambda = 0.70 \rightarrow \lambda = 0.80$	−1.017	$\lambda = 0.70 \leftarrow \lambda = 0.80$	1.004
$\lambda = 0.80 \rightarrow \lambda = 0.90$	−1.309	$\lambda = 0.80 \leftarrow \lambda = 0.90$	1.231
$\lambda = 0.90 \rightarrow \lambda = 1.00$	−1.495	$\lambda = 0.90 \leftarrow \lambda = 1.00$	1.582
Total:	−6.449	Total:	6.514

As can be seen, the free energy difference calculated in the forward and reverse directions is almost equal in magnitude (though of course opposite in sign), with a difference of less than 0.07 kcal mol^{-1}. For each of the steps, the forward and reverse free energy changes likewise are consistent to within better than 0.15 kcal mol^{-1}. These differences provide one measure of the numerical error in the procedure (several other statistical measures of the error are available but go beyond the scope of this book), and their small magnitude suggests that the free energy perturbation approach has yielded reliable results. The quality of the results also depends on the forcefield parameters used for water, ethane, and methanol. In this case, these appear to be of sufficient quality, because the experimental difference in free energy of solvation is 6.94 kcal mol^{-1} (the experimental free energies of solvation are +1.83 for ethane and −5.11 for methanol), with which the calculated value is in good agreement. The results are also shown graphically in Figure 7.3, which shows that the bulk of the difference in free energy of solvation occurs between $\lambda = 0.5$ and 1, i.e. when the system is closer in properties to methanol than to ethane.

Free energy perturbation can be applied to many different problems. One popular application is the computation of the relative free energy of interaction between two molecules C and D and an environment, for example a solvent. Application of equation (7.25) yields the difference in the free energy of solvation of the two species. Another example is where the environment is the active site of a protein: in this case you obtain the difference in the free energy of binding of the isolated, gas-phase molecules to the protein. Combined with a separate calculation of the relative free energy of solvation, this yields the difference in the free energy of binding of the two molecules from solution. The two molecules do not need to contain the same atoms— for example, C may have a –CH$_3$ substituent, while D has an –F substituent. This is a so-called 'alchemical' transformation.

Many other methods are available for computing free energies based on molecular dynamics simulations and enhanced sampling methods. Indeed, almost all problems

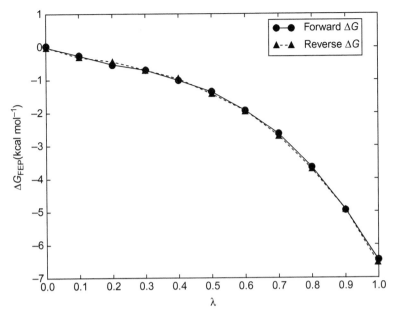

Figure 7.3 Free energy changes calculated in a free energy perturbation calculation for the relative solvation free energy of ethane ($\lambda = 0$) and methanol ($\lambda = 1$). The energy plotted at each value of λ corresponds to the sum of partial free energies obtained for each step between 0 and that value of λ.

in which the aim is to compute relative free energies based on simulations face the 'rare event' challenge, in that the desired transformation is unlikely to occur frequently enough in a spontaneous way even in a very long simulation, so that the 'direct' approach used to generate the data in Figure 7.2 is not applicable. Here we have given a long overview of the free energy perturbation approach so as to give an idea of the detailed working of one particular approach. This is not meant to suggest that free energy perturbation is the only possible solution. Other techniques such as metadynamics, umbrella sampling, thermodynamic integration, and many others also play an important role. Concerning problems, we have emphasized here differential solvation, and binding to proteins. Many other systems can be tackled using enhanced sampling techniques, and the ideal approach to be used will vary depending on the problem.

7.6 **Further reading**

As mentioned in Chapter 6, the Further reading sections for both chapters overlap to some extent.

- *Molecular Driving Forces: Statistical Thermodynamics in Biology, Chemistry, Physics and Nanoscience*, 2nd Edition, Ken A. Dill and Sarina Bromberg, Garland Science, London and New York, 2010. The work in this chapter builds on a foundation of statistical mechanics. There are many excellent textbooks on statistical mechanics, approaching the subject from diverse viewpoints. Somewhat unusually for the field, this book focuses on concepts rather than extensive mathematics.

- *Understanding Molecular Simulation, From Algorithms to Simulation*, 2nd Edition, Daan Frenkel and Berend Smit, Academic Press, San Diego, 2002. This classic book on molecular simulations describes algorithms, statistical mechanics, and the approach to macroscopic quantities from computation.

- Good Practices in Free-Energy Calculations, Andrew Pohorille, Christopher Jarzynski, and Christophe Chipot, *Journal of Physical Chemistry B*, 2010, **114**, 10235–53. Extracting free energies from simulations is in some ways an art form as well as a science, and in-depth analysis of the sources of error is important to make headway. This review article articulates many of the key aspects of this analysis.

7.7 Exercises

7.1 Use the procedure in section 7.2 to calculate the equilibrium constant for a simple gas-phase reaction, such as $SO_2 + \frac{1}{2} O_2 \rightarrow SO_3$. Use a simple level of quantum mechanical treatment such as HF/6-31G to optimize the structures and compute the vibrational frequencies, and use the statistical mechanics output of the quantum chemistry code. Recalculate the reaction energy (which is equal to ΔG (0 K)) using a more accurate method, such as MP2/6-31G(d), and compare both values of ΔG^0 obtained to experiment.

7.2 Using the approach described in section 7.3, calculate the rate constant for a simple gas-phase reaction, such as addition of ozone to ethene to yield a cyclic ozonide. Again, use a simple quantum chemical method such as B3LYP/6-31G(d) to compute the structure and vibrational frequencies of the reactants and the TS. Compare your predicted rate constant to the experimental value. Assess the sources of error.

7.3 Some MD simulation codes contain in-built tools to carry out biased simulations and to extract free energy profiles from them. Umbrella sampling is a technique that is often available. Use umbrella sampling to compute the potential of mean force for *cis–trans* isomerization of the peptide bond in a small peptide dissolved in water.

7.8 Summary

- Statistical thermodynamics can be used to calculate thermodynamic properties such as relative free energies based on properties of the potential energy surface $V(R)$.
- For gas-phase molecules, this can be done by evaluating the quantum mechanical molecular partition function in terms of the translational, rotational, and vibrational properties of the molecules.
- Combined with accurate quantum chemical calculations, this can yield very accurate free energies.

- In a similar way, activation free energies as used in transition state theory can be obtained; again, accurate prediction of rate constants is possible for gas-phase reactions when reliable quantum chemical data is used.
- Statistical thermodynamics for extended systems relies on properties of the classical partition function for the system.
- While this cannot be explicitly evaluated, it serves as a basis for both direct and biased methods for free energy evaluation; a popular biased method is referred to as the free energy perturbation method, which is used extensively in studying ligand binding to biomolecules.

Hybrid and Multi-Scale Methods

8.1 Introduction

The previous three chapters have given an overview of how potential energy surfaces can be explored, using static and dynamical methods, culminating in methods to predict free energy differences that can be related to experiment. In this chapter, the focus returns to methods for calculating potential energy surfaces. Chapters 2, 3, and 4 provided an introduction to quantum mechanical and molecular mechanical techniques for computing the energy for a given structure. With these methods, the whole system to be treated is described using the same theoretical approach. Molecular mechanical approaches are ideal when changes in electronic structure are not involved, but they are not suited for describing the many problems in chemistry that involve bond-making and -breaking. On the other hand, quantum chemical methods cope well with changes in bonding, but are limited to describing relatively small systems, due to the high scaling of the computational effort with system size, so they appear at first sight to be limited to describing gas-phase chemistry of isolated small molecules. Periodic quantum chemical approaches can, however, be used to describe crystalline systems with small unit cells.

When considering chemistry of more complex extended and condensed-phase systems being restricted to finite models of modest sizes appears to be a serious limitation. The methods described in the present chapter have been developed in order to overcome this problem. Each of the methods described here are based on combining detailed quantum chemical description of bonding in a smaller core region with a simpler description of the environment and of its effects on the core region. Methods to describe systems that are too large even for treatment with molecular mechanics are also briefly described.

Before describing these methods it is useful to pause and consider to what extent quantum chemical methods really are unsuited to describe anything other than gas-phase processes for small molecules. This turns out only to be true for cases where the effect of the environment on the system is large. Where it is not, reasonable results will be obtained from quantum chemical studies in which the surroundings are simply ignored. Many properties, whether spectroscopic, structural, mechanistic, or thermodynamic, rely ultimately on *differences* in energy between two states. The environment will influence the property by changing the energy difference, by interacting more strongly with one of the two states than with the other. Quite often, though, a physical analysis of the problem will show that both states should interact roughly equally with the environment, so that the effect of the latter on the target property is small.

Consider for example the NMR spectrum of a molecule in solution, where the molecule (and applied magnetic field) is the 'core' and the solvent the environment, and the spectrum depends on the difference in energy between different nuclear spin arrangements. The effect of the solvent on this difference will often be tiny, so the effect of solvent can be neglected. Another example is reaction of neutral compounds involving non-polar intermediates and TSs, where the solvent will interact roughly equally with the reactants and the TS, thereby having little effect on the rate constant. Where the comparison can be made, such reactions often have roughly the same rate constant in the gas phase and in solution (once any necessary unit conversion has been carried out). For example, the Diels–Alder reaction of cyclopentadiene with itself to make a dimer has a rate constant in tetrachloromethane solvent that is the same, within a factor of 2, as the rate constant in the gas phase.

Where interactions with the environment are stronger, they can sometimes be treated in quantum chemical calculations by including them as part of the system. A sidechain group in an enzyme that participates in a catalytic mechanism can simply be included in the reactive system; likewise the few solvent molecules that interact strongly with a solute. At some point, though, there will be strong enough interactions with a large enough environment that they simply must be treated, and that is where the methods of this chapter will need to be used.

8.2 Continuum models of the environment

At large enough length-scales, matter can be described by macroscopic properties such as density, viscosity, and dielectric permittivity. At the molecular level, such a description is no longer applicable, but, in many cases, such a bulk description of the environment around a molecular system can provide a remarkably accurate model of phenomena such as solvation. Continuum models of solvation can be used in combination with both quantum chemical and molecular mechanical methods. The goal of such methods is to calculate the standard free energy of solvation $\Delta_{solv}G$ for the system being described. The solvation free energy can be defined in a number of different ways depending on which standard reference state is chosen for the gas phase species and the solute. Continuum models almost always set out to calculate the standard free energy change between the ideal gas and the solute with each taken at the same overall reference concentration.

One of the simplest continuum models for solvation is the so-called *Born* model of equation (8.1), which applies to spherical solutes with a charge q placed at the centre of the sphere. It can be obtained by integration of the Poisson equation of classical electrostatics, which itself rests on the assumption that the solvent cage polarizes linearly in response to the charge on the solute.

$$\Delta_{solv}G = \frac{-1}{2}\left(1 - \frac{1}{\varepsilon_r}\right)\frac{1}{4\pi\varepsilon_0} \times \frac{q^2}{r} \qquad (8.1)$$

In the equation, ε_0 is the vacuum permittivity, ε_r is the relative permittivity of the solvent, and r is the radius of the solvation sphere or cavity created by the solute. As can be seen, the equation bears a strong similarity to Coulomb's Law, which reflects the fact that the Born model essentially sets out to describe the coulombic interactions between the solute and a field of partial charges created in the continuum. Consider a chloride ion Cl^-, using for r the van der Waals radius of chlorine, 1.75 Å. Taking $\varepsilon_0 = 8.85419 \times 10^{-12}$ C^2 m^{-1} J^{-1}, $\varepsilon_r = 80$ for water, $q =$ the charge of an electron $= 1.6022 \times 10^{-19}$ C, one obtains $\Delta_{solv}G$ for one ion of -6.51×10^{-19} J, or -392 kJ mol^{-1}, vs the experimental value of -304 kJ mol^{-1}. The model clearly gives the correct magnitude for the solvation free energy, showing that the use of bulk or continuum properties for the solvent is reasonable. The difference between the computed and experimental values may be viewed as being due to the value assigned to the radius, r, or to a breakdown of the assumption concerning linear polarization response.

Because of the simple nature of the spherical cavity, integration of the Poisson equation can be carried out analytically to yield the Born equation with its simple expression for the solvation free energy. If one instead resorts to a numerical treatment of the Poisson equation, the cavity shape can be much more complex, and multiple partial positive and negative charges can be distributed around the cavity. In such a framework, it also becomes possible to account for non-electrostatic interactions between the solute and the solvent, such as dispersion. Finally, one can include terms that account for the free energy needed to create a cavity in the bulk solvent. This yields a *polarizable continuum model* (or PCM) of the solute–solvent interaction.

PCM models can vary depending on the way in which the cavity is defined, or on whether or not non-electrostatic interactions are included. Also, they have a number of parameters. Hence there are many different PCM models, with a variety of different names, including COSMO (conductor-like screening model), SMD (screening model based on density), and so on. These models are widely available in quantum chemical and molecular mechanics codes. For quantum chemical implementations, the electrostatic part of the interaction is usually treated self-consistently. This means that instead of minimizing the energy of the wavefunction or density in the Hartree–Fock or Kohn–Sham procedure, one minimizes the overall energy including the electrostatic interaction with the 'reaction field' of partial charges in the continuum solvent created in the solvent in the vicinity of the cavity. Such models are also known as *self-consistent reaction field* (SCRF) models. PCM or SCRF approaches are illustrated in cartoon form in Figure 8.1.

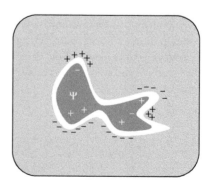

Figure 8.1 Cartoon showing the outcome of an SCRF calculation of a solute–solvent interaction: the partial charges of the wavefunction Ψ of the solute induce charges in the solvent, treated as a continuum, and these interactions are treated self-consistently.

Continuum solvent models provide a good qualitative and even quantitative description of solvation effects. For a test set of free energies of solvation of neutral species in several common solvents including water, the commonly used SMD continuum model yields a mean error relative to experiment of 0.59 kcal mol^{-1}. It is to be noted that the experimental values are all fairly close to zero, so while the errors are small in absolute terms, they remain sizeable in relative terms. For ionic solutes, which are much more strongly solvated, especially in polar solvents, the free energies of solvation are all negative and much larger in magnitude. Here too, good accuracy is achieved, with mean errors of 3.20 kcal mol^{-1} for ionic solutes in water. Other continuum models yield similar levels of accuracy.

Continuum models can be used in combination with quantum chemical methods in order to predict properties that are strongly influenced by solvent. Here we consider two examples: acid–base equilibria, and reactivity of charged species.

Dissociation of a weak acid HA in a solvent, such as water, is associated with an equilibrium constant called the acid dissociation constant K_A, equation (8.2):

$$HA \rightleftharpoons A^- + H^+ \; ; K_A = [H^+][A^-]/[HA] \tag{8.2}$$

The dissociated proton H$^+$ is not in reality present as a free species, as it interacts quite strongly with water. This can be represented by including a water molecule in the reaction, and writing the product as H_3O^+ (though even this understates the degree of solvation of the proton), yielding equation (8.3):

$$HA + H_2O \rightleftharpoons A^- + H_3O^+ \; ; K'_A = [H_3O^+][A^-]/[HA][H_2O] = K_A/[H_2O] \tag{8.3}$$

Using this second way of writing the reaction, we can set up a thermodynamic cycle (Figure 8.2) relating the standard free energy change of the reaction in the gas phase $\Delta G^0_{(g)}$, the standard free energies of solvation of the different species $\Delta_{solv}G$, and the standard free energy change of the reaction in water $\Delta G^0_{(aq)}$. The gas-phase free energy change can be computed using the methods of Chapter 7, and SCRF methods can be used to predict the free energies of solvation. One then obtains equation (8.4) for the free energy in solution:

$$\Delta G^0_{(aq)} = \Delta G^0_{(g)} + \Delta_{solv}G(A^-) + \Delta_{solv}G(H_3O^+) - \{\Delta_{solv}G(HA) + \Delta_{solv}G(H_2O)\} \tag{8.4}$$

And from this, one can calculate the pK$_A$ of the acid, equation (8.5):

$$pK_A = -\log_{10}K_A = \frac{-\Delta G^0_{(aq)}}{RT\ln 10} \tag{8.5}$$

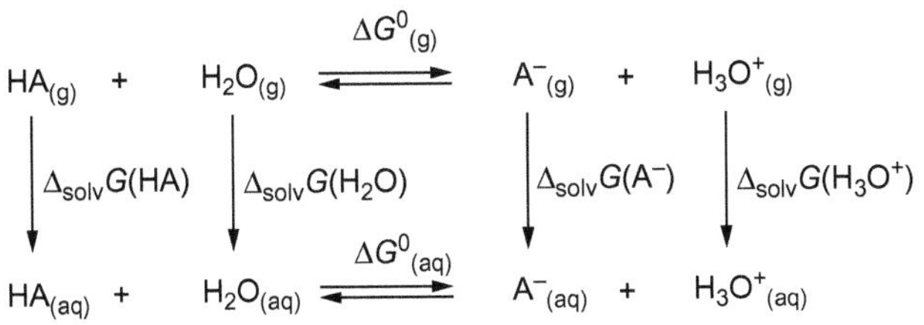

Figure 8.2 Free energy cycle for calculating the pK$_A$ of a weak acid in water.

Ab initio prediction of pK_A for a weak acid

Once again we will illustrate the potential of this approach using examples, the pK_As of the weak organic acids: formic acid (HCOOH) and acetic acid (CH$_3$COOH). Gas-phase geometry optimization has been performed using DFT with the B3LYP functional (including a dispersion correction referred to as D3BJ) and the 6-31G(d) basis set. Vibrational frequencies at the same level of theory have been computed for all species, allowing corrections for zero-point energy to be included, and calculation of the free energy at $T = 298.15$ K, using the methods of Chapter 7. It is to be noted that the program calculates the free energy terms assuming an ideal gas standard state $p^0 = 1$ atm (i.e. a molar volume of 24.5 dm^3). This has been corrected to a solution phase standard state of $c^0 = 1$ M (i.e. a molar volume of 1 dm^3) for all species other than water, by adding a term $RT \ln (24.5/1)$ to all free energy corrections. For water, the standard state used is the pure liquid ($c^0 = 55.5$ M so $V_m = 1/55.5$) so the correction term used was instead $RT \ln (24.5 \times 55.5)$. The energy at the optimized structure was recalculated using B3LYP, still using the D3BJ dispersion correction, and with a much larger basis set, 6-311++G(2d,p). The free energy of solvation of the species was calculated with the SMD continuum solvent model at the B3LYP/6-31G(d) level. Table 8.1 shows the calculated data.

It can be seen that the reaction in equation (8.3) is extremely unfavourable in the gas phase, with an energy change of over 700 kJ mol^{-1} for both acetic acid and formic acid. There is a notable change of over 30 kJ mol^{-1} between the small basis set and the larger one. The difference between the gas phase potential energy and the free energy, $\Delta G - \Delta G(0)$ (which includes the zero-point energy correction in Table 8.1) does not make a large change to the overall calculated energy. The free energies of solvation for the neutral acids and for water are relatively small, but for the hydronium cation and the anionic conjugate bases, these values are in excess of 200 kJ mol^{-1}, so this term makes a large contribution to the overall calculated free energy. From the data in Table 8.1, it is possible to calculate the pK_A for the two acids. The overall free energy changes of equation (8.4) are of 101.3 and 82.5 kJ mol^{-1}, which, together with equation (8.5), yields predicted pK_A values of 17.75 and 14.47 for the two acids, versus experimental values of 4.75 and 3.75, respectively. The large errors in the predicted values, of over 10 pK_A units, correspond to errors in the free energy changes of over 60 kJ mol^{-1}. What is the source of this error? There are certainly contributions

Table 8.1. Data for calculation of the pK_A of acetic acid CH$_3$COOH and formic acid HCOOH. All energies in hartree, except for the calculated changes for equation (8.3), which are in kJ mol^{-1}.

	B3LYP 6-31G(d)	$\Delta G - \Delta G(0)$	$\Delta_{solv} G$	B3LYP 6-311++G(2d,p)	$\Delta G^0_{(aq)}$
H$_2$O	−76.40953	0.01031	−0.01313	−76.46006	−76.46288
H$_3$O$^+$	−76.69015	0.01921	−0.15320	−76.73274	−76.86674
CH$_3$COOH	−229.08879	0.03796	−0.00871	−229.17700	−229.14775
CH$_3$COO$^-$	−228.50409	0.02388	−0.11608	−228.61312	−228.70532
Δ eq (8.3)	798.3	−13.6	−649.6	764.5	101.3
HCOOH	−189.75854	0.01287	−0.00783	−189.83489	−189.82985
HCO$_2^-$	−189.17993	−0.00053	−0.11570	−189.27832	−189.39456
Δ eq (8.3)	782.4	−11.8	−651.0	745.3	82.5

(continued...)

both from the electronic energy term due to inaccuracies in the DFT method used, and the basis set used. Likewise, the $\Delta G - \Delta G(0)$ term is not perfectly accurate. However, the dominant source of error here is the free energy of solvation, $\Delta_{solv}G$, and indeed, more specifically, the calculated free energy of solvation of H_3O^+. As already mentioned, this small ion interacts very strongly with surrounding water, and even the accurate SMD model struggles to reproduce the free energy of solvation accurately.

Fortunately, for calculating pK_A, another approach is possible that side-steps the problems with calculating the pK_A 'directly' through use of equations (8.4) and (8.5). The free energy of the proton in the gas phase is known and so is the experimental free energy of solvation of the proton. Together these give a solution phase free energy for the proton at standard state 1 M of −0.4293 hartree. Combined with the other data in Table 8.1, and equations (8.2) and (8.5), this leads to predicted pK_As of 6.05 and 2.77 for the two weak acids, much closer to experiment. The remaining errors are of the order of 2 pK_A units, which is equivalent to about 10 kJ mol^{-1}, a magnitude of error that should be expected in any *ab initio* procedure.

Continuum methods can also be used to assess the effect of solvation on reactivity. The thermodynamic cycle of Figure 8.3 can be used to suggest an expression for the free energy of activation in solution. Based on this diagram, the following expression, equation (8.6) can be obtained for the free energy of activation of reaction in solution:

$$\Delta G^{\ddagger}_{(sol)} = \Delta G^{\ddagger}_{(g)} + \Delta_{solv}G(TS) - \{\Delta_{solv}G(A) + \Delta_{solv}G(B)\} \tag{8.6}$$

In practice, a number of different approaches can be used to evaluate equation (8.6). One option is to calculate the free energy of activation of the reaction as accurately as possible in the gas phase, then to compute the free energy of solvation of each of the reactants and the TS using the continuum model, without modifying the structure. This approach is frequently successful, but it neglects an occasionally important effect of solvation: species sometimes have a rather different structure in solution than in the gas phase. Hence computation of $\Delta_{solv}G$ for the gas-phase structure may not be ideal. The change of structure may also affect the vibrational frequencies and rotational constants of the different species, so that the free energy of activation calculated from statistical thermodynamics using the gas-phase properties may need to be corrected. This problem can be side-stepped by including the continuum solvent already in the geometry optimization and vibrational frequency computations. In this way, the effect

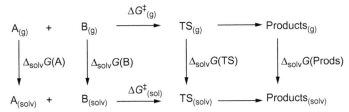

Figure 8.3 Free energy cycle for calculating $\Delta G^{\ddagger}_{(sol)}$, the free energy of activation of a reaction in solution.

of solvation is already treated in the statistical thermodynamics. This approach can be particularly important for reactions, as solvation effects can strongly perturb the structure of TSs. However, the same points can also be made for stable species and intermediates, so use of the continuum model can be useful also for calculating equilibrium thermodynamics, as e.g. in the pK_A example given above.

Gas-phase vs solution phase S_N2 reactions

Here we apply equation (8.6) in the case of a simple chemical reaction displaying a large dependence on solvent, namely the S_N2 substitution $Cl^- + CH_3Br \rightarrow Br^- + CH_3Cl$. The kinetics of this reaction have been measured in the gas-phase ($k_g = 1.3 \times 10^{10}$ M^{-1} s^{-1}) as well as in several solvents, including acetone ($k_{solv} = 3.3$ M^{-1} s^{-1}) and water ($k_{aq} = 5.0 \times 10^{-6}$ M^{-1} s^{-1}). As can be seen, there is a very strong dependence on solvent polarity, with the reaction occurring over 10^{15} times slower in water than in the gas phase. This can be understood based on the free energy diagram in Figure 8.4. This shows that in the gas phase there is a *minimum* corresponding to a 'complex' between the chloride ion and methyl bromide. In the gas phase, there is a strong attraction between the dipole moment on the methyl bromide and the negative charge on chlorine. Likewise, the complex $ClCH_3 \bullet Br^-$ is more stable than separate CH_3Cl and Br^-.

A further notable aspect of Figure 8.4 is that the complexes and the TS are less strongly stabilized by solvation than the separate reactants. This can be understood based on equation (8.1): isolated Cl^- and Br^- have a full negative charge, and are fairly compact, hence they have a relatively small radius r. The complexes and TS have somewhat delocalized negative charge, and also have effective radii that are greater than those of the isolated anions. Charge delocalization leads to smaller solvation free energy, due to the q^2 dependence in equation (8.1): two charges of -0.5, as in the TS, are less favourable than one charge of -1. Increased radius leads to a less close approach by the solvent dipoles, so also causes lower solvation free energy. Because of this second aspect, the free energy of the TS relative to reactants is much higher in solution than in the gas phase—and this effect should increase with increasing polarity of the solvent, as is observed experimentally. Also, the binding energy of the complexes relative to reactants or products is weakened in solution, and indeed in polar solvents such complexes are not even minima.

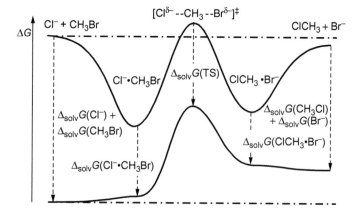

Figure 8.4 Qualitative free energy surfaces for the $Cl^- + CH_3Br$ reaction in the gas phase and in solution.

(continued...)

Table 8.2 Gas phase and solution free energies relative to reactants (in kJ mol^{-1}) for the $Cl^- + CH_3Br$ system, using gas-phase ωB97X-D/6-31+G(d) structures and statistical thermodynamics free energies (with a 1 M standard state and $T = 298.15$ K), together with free energies of solvation from the SMD continuum solvent model with parameters for acetone.

Species	$\Delta G_{(g)}$	$\Delta G_{(solution)}$
$Cl^- + CH_3Br$	0.0	0.0
$Cl^- \bullet CH_3Br$	−48.8	10.9
TS	13.3	100.7
$ClCH_3 \bullet Br^-$	−34.9	38.5
$ClCH_3 + Br^-$	−18.7	31.0

How well do calculations with and without a continuum solvent model reproduce these effects, qualitatively and quantitatively? The outcome of calculations using the ωB97X-D DFT functional, the 6-31+G(d) basis set, and the SMD solvation model is summarized in Table 8.2. It can be seen that the qualitative picture from Figure 8.4 is reproduced, with the minima for the complexes being lower than the separated species in the gas phase, but *higher* in solution, and a free energy barrier to reaction that is much larger in solution than in the gas phase. Taking the free energy barriers in the gas phase and in solution, the Eyring equation predicts rate constants of $k_g^{calc} = 2.85 \times 10^{10}$ M^{-1} s^{-1} and $k_{solv}^{calc} = 1.4 \times 10^{-5}$ M^{-1} s^{-1}, respectively 2.3 times larger than experiment and 2.4×10^5 times smaller. The quantitative agreement with experiment is less good in solution than in the gas phase, presumably due in large part to deficiencies in the solvent model. (It should be noted that a factor of 2.4×10^5 error can be caused by an error in free energy of 31 kJ mol^{-1}, not much larger than the mean error for SMD-calculated free energies of solvation of ionic species of 3.2 kcal mol^{-1} or 13 kJ mol^{-1} mentioned above.)

8.3 Hybrid methods

For large systems, it would be attractive to be able to perform calculations treating different parts of the model using different methods, so as to treat each part with the correct combination of accuracy and computational efficiency. For example, you might wish to obtain a very detailed and accurate description of a core region but be prepared to accept a much lower accuracy for the environment and its coupling to the core. The continuum environment models of the previous section are one example of such a hybrid approach. Here, the contrast in level between the treatment of the two regions is extreme, with a quantum chemical treatment of the core, and a bulk, continuum description of the environment. Methods in which both regions are treated with atomistic detail but with different levels of theory can also be put forward and implemented, and these are referred to as *hybrid* models.

There are a large variety of such models, usually with the core region treated using some or other quantum chemical or quantum mechanical method, and the environment treated using simpler molecular mechanical approaches. Such methods are often referred to as 'QM/MM' methods. There are also other methods in which both regions are treated with quantum chemistry, but with different levels of theory ('QM/QM' methods). It is also possible to generalize to methods with a core, a first environment, and a second environment (QM/QM/MM, and so on). For the two-level case with a high-level (HL) method for the core, and low-level (LL) for the environment, the overall potential energy expression for a given set of coordinates of the atoms in the core (R_C) and environment (R_E) can be written as:

$$V^{HL/LL}(R_C,R_E) = V^{HL}(R_C) + V^{LL}(R_E) + V^{HL-LL}(R_C,R_E) \qquad (8.7)$$

The first and second terms on the right-hand side of equation (8.7) can be computed using standard methods. For example if HL is a quantum-chemical technique, the first term is simply the same as the methods described in Chapters 2 or 3. Likewise for MM as the low level, the second term is that described in Chapter 4. The third term describes the interactions between the two regions, and defines the type of hybrid method that is used.

Hybrid methods are simpler if there is no covalent bonding nor strong charge-transfer present between the core and the environment, so we will describe this case first. For both QM/QM and QM/MM couplings, one popular approach, developed by Japanese theoretical chemist Keiji Morokuma, is referred to using the 'ONIOM' acronym (which stands for 'Our own N-layered Integrated molecular Orbital and molecular Mechanics method'), and uses the following energy expression, equation (8.8):

$$V_{ONIOM}^{HL/LL}(R_C,R_E) = V^{HL}(R_C) + V^{LL}(R_C,R_E) - V^{LL}(R_C) \qquad (8.8)$$

For example, consider a complex between a chloride ion and water, $Cl^- \bullet H_2O$ (Figure 8.5) to be described with an ONIOM type dual-level energy using MP2/6-311+G(d) as the high level and HF/6-31+G as the low level. This requires two calculations on the isolated chloride ion, one with MP2 and the 6-311+G(d) basis (returning an energy of −459.70357 hartree), and the other with HF and the 6-31+G basis (returning −459.53828 hartree), and one on the whole complex with HF/6-31+G (−535.54824 hartree). The overall MP2/6-311+G(d):HF/6-31+G energy is then −459.70357 + −535.54824 − (−459.53828) = −535.71353. In itself, this provides no additional insight, but the *gradient* of the energy is also the sum of the MP2 gradient on the core and the HF gradient on the whole system, minus the HF gradient on the core. This provides a means to optimize geometries using the hybrid method, which is useful. Also, one can compare energies computed for different species at the same hybrid level of theory, obtaining results that reflect the structure of the whole system.

Figure 8.5 Simple chloride anion (core)–water molecule (environment) system.

With the simple ONIOM procedure, the coupling between the core and the environment is described only at the lower level. More sophisticated QM/QM methods, sometimes referred to as 'embedding' methods, allow the coupling to be described in a more accurate way.

For systems where there are covalent bonds between the core and environment, a QM calculation on the core atoms alone will not be meaningful as there will be one or more unpaired electrons on the atoms at the junction, inducing an artificial electronic structure at this position. In such cases, the QM calculation on the core needs to include some extra 'capping' feature. The simplest procedure to treat the 'link' between the core and environment is to introduce additional 'link' atoms, usually hydrogen atoms, to saturate the valences. In the example of the ruthenium complex $RuCl_2(P(C_6H_{11})_3)_2(CHC_6H_5)$ shown in Figure 8.6, for example, the central ruthenium atom, as well as the two phosphorus and two chlorine atoms directly bonded to it, and the CH atoms of the phenylcarbene ligand, could form the core. These atoms are connected by seven covalent bonds to the surrounding cyclohexyl and phenyl groups, and these groups would need to be replaced by 'link' hydrogen atoms in a hybrid QM/MM calculation.

Hybrid QM/MM methods are frequently used to describe biomolecular systems, usually using a different coupling scheme called 'electrostatic embedding' in equation (8.7). Here the aim is to gain insight into the electrostatic effects of the polar groups in the protein environment on the electronic structure of the core or QM region, typically composed of the atoms of the active site and the substrate in an enzyme. One

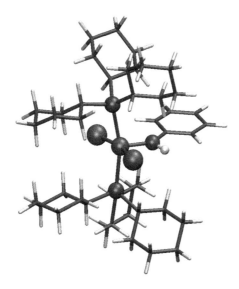

Figure 8.6 Example of a large organometallic complex, $RuCl_2(P(C_6H_{11})_3)_2(CHC_6H_5)$, containing 120 atoms. This model could be described by a hybrid method where the seven key atoms, $\{RuCl_2P_2CH\}$, shown as spheres, could be treated at the high level, with the surrounding C_6H_{11} and C_6H_5 groups, shown in stick form, being treated at the low level. In this case, seven link atoms would typically also need to be used. An ONIOM-type model with coupling described only at the lower level would still account for the important 'mechanical' effects of the environment on the core.

efficient way to describe this coupling is to include the electric field created by the MM point charges of the environment (the 'MM region') in the quantum chemical electronic Hamiltonian of the core region, yielding equation (8.9):

$$\hat{H}_{elec}^{QM/MM} = \frac{-1}{2}\sum_{i=1}^{n}\nabla_i^2 + \sum_{i=1}^{n}\sum_{A=1}^{N}\frac{-Z_A}{r_{iA}} + \sum_{i=1}^{n}\sum_{j>i}^{n}\frac{1}{r_{ij}} + \sum_{i=1}^{n}\sum_{K=1}^{N_{MM}}\frac{-q_K}{r_{iK}} \tag{8.9}$$

This expression is identical to equation (2.2), except for the fourth and final term on the right-hand side, accounting for the coulombic interactions between the MM point charges q_K and the electrons of the core region. As can be seen, this term is of the same form as the second term (the electron–nucleus interaction term), and can be readily evaluated in quantum chemical codes. Also, evaluating this term adds little to the computational challenge, since the electron–electron part of equation (8.9) is usually more demanding than the other parts.

This type of QM/MM approach allows polarization of the core region by the MM point charges. For example, in the chloride ion–water system of Figure 8.5, the point charges associated with the O and H atoms of the water molecule can polarize the electron cloud of the chloride anion. However, such calculations do not describe the associated polarization of the MM region by the QM region. As mentioned in Chapter 4, variants of MM with polarizable atoms do exist, and can also be used in more sophisticated QM/MM hybrid methods in which the reverse polarization is then included.

QM/MM methods with either ONIOM-type coupling or electrostatic embedding as in equation (8.9) have become very popular as they allow the quantum chemical modelling of very large systems at relatively modest computational expense. These methods remain, however, typically much more demanding than MM. As discussed in Chapters 5, 6, and 7, effective exploration of the potential energy surface and calculation of free energies for very large systems typically requires *simulation* methods rather than geometry optimization techniques, due to the large number of local conformational minima. This aspect requires careful attention when performing QM/MM calculations, since these often do not allow simulations given the expense of the QM part of the calculation.

8.4 Coarse-grained molecular mechanics models

The molecular mechanics methods of Chapter 4 are relatively undemanding in computational terms, at least compared to quantum chemical methods. Nevertheless, for very large systems, the computational expense of MM can prove unaffordable. In such cases, it can be useful to perform modelling using the same type of formalism as MM, but using particles larger than single atoms as building blocks. For example, a methyl group $-CH_3$ could be treated as a single particle, with mass 15 amu, and appropriate steric properties. Even larger groups can be used, e.g. the isopropyl sidechain $-CH(CH_3)_2$ of the amino acid valine could be treated as a single particle. With careful choice of the parameters for such coarse-grained models, it is possible to reproduce fairly accurately the properties that would be obtained in a full atomistic model.

There are at least two main advantages of coarse-grained models. First, they reduce the number of coordinates that need to be treated—replacing methyl or isopropyl by one single entity removes at one stroke nine or even twenty-seven degrees of freedom. While the remaining particle may require a slightly more complex energy expression than a single atom in an atomistic model, the saving in terms of computational effort of carrying out this substitution can be very substantial for a large system. The second advantage is that in most coarse-grained models the lightest particles are significantly heavier than the hydrogen atoms present in atomistic models. This also means that the highest frequency motions present in the system have considerably longer periods than X–H bonds. This allows the use of larger time-steps in molecular dynamics studies, so that it is possible to simulate longer total times within the same number of time-steps. Of course, it is also impossible in such a model to gain any insight into the internal motions of atoms within the coarse grains, e.g. the vibrations of the C–H bonds in the methyl groups.

Coarse-grained models can be combined with molecular mechanics or even quantum chemistry to yield hybrid models similar to those of section 8.3. As for coarse-grained models in general, the difficulty for such types of calculation lies in generating the parameters accurately.

8.5 Further reading

- Quantum mechanical continuum solvation models, Jacobo Tomasi, Benedetta Mennucci and Roberto Cammi, *Chemical Reviews* 2005, **105**, 2999–3093. This monumental review—longer than the present book!—provides a general overview of research on continuum solvent models.

- Theoretical studies of enzymic reactions: dielectric, electrostatic and steric stabilization of the carbonium ion in the reaction of lysozyme, Arieh Warshel and Michael Levitt, *Journal of Molecular Biology* 1976, **103**, 227–249. This is generally recognized to be the first paper describing QM/MM calculations. It also provides an excellent introduction to the value and nature of such methods.

- The Nobel Prize in Chemistry 2013—Advanced Information. Nobelprize.org, Nobel Media AB 2014. http://www.nobelprize.org/nobel_prizes/chemistry/laureates/2013/advanced.html, retrieved on 16 Jul 2017.

8.6 Exercises

8.1 Use a quantum chemical code and a simple level of theory (e.g. HF/6-31G) to compute the solvation free energy of a neutral molecule (e.g. methanol) and an ion (e.g. methoxide) in a non-polar solvent (such as benzene), a polar aprotic solvent (such as THF), and a protic solvent (such as methanol). Compare the values obtained and rationalize them based on solvent–solute interactions.

8.2 Using the ONIOM or related hybrid method, calculate the bond energy for a sterically hindered system such as the central C–C bond in hexaphenylethane, and compare to the bond energy for ethane itself. Examine the contribution to the bond energy from the MM part of the molecule. For background, see Stefan Grimme and Peter Schreiner, *Angewandte Chemie International Edition*, 2011, **50**, 12639–42.

8.7 **Summary**

- Chemistry of a 'core' system is frequently perturbed by the 'environment'.
- If the perturbation is weak, as is often the case, it is acceptable to perform calculations on the core part only.
- The effects of the environment can in many cases be described using continuum models, that can conveniently be coupled to quantum chemical calculations.
- It is also possible to devise hybrid methods, in which the atoms making up the core and the environment in the model are treated at different levels of theory.
- One very popular family of hybrid methods treats the core quantum mechanically (QM) and the environment with molecular mechanics (MM), and these methods are referred to as QM/MM methods.

9 Conclusions

9.1 Introduction

Computational methods are used in a wide variety of ways to provide a better under-standing of chemical phenomena. The theoretical description of matter based on the basic physics of atoms and molecules might seem to be an impossibly challenging task, given the enormous complexity of quantum mechanics, and indeed, except for the very simplest systems, it is necessary to use approximate methods in any practical modelling of chemistry. However, computers are more and more powerful, new and more accurate approximations are being developed, and algorithms are becoming more and more efficient, so that overall it is now easier than ever before to perform calculations that provide accurate models of experimental systems in all areas of chemistry. These developments are expected to continue in coming years, making this a golden age for computational chemistry.

This book has provided a first introduction to many of the commonly used methods and techniques of computational chemistry. Given the constraints of space, the book is very far from providing a comprehensive introduction to every aspect of every method. Nevertheless, it is hoped that the descriptions provided form a basis for use of the methods in research, and for further study.

An important focus of the book is to provide results of genuine applications and discuss some of the practical issues that arise when performing such calculations. The examples selected—all based on calculations performed by the author in writing the book—were chosen based on their simplicity, but the issues touched upon in the context of these examples are frequently of broader relevance. Also, the calculations described cover a significant part of the wide range of methods and methodological choices that are made in contemporary research studies. In the next section, an attempt is made to draw together some general aspects that should be borne in mind when designing a computational study.

9.2 Designing a computational project

As for any scientific project, research in computational chemistry needs some degree of planning in order to be successful. The whole of this book is aimed at helping to make the choices involved in such planning. In the summary at the end of each chapter, the exercises, and the many worked examples, readers should have found many hints aimed at assisting with project design. It is however useful to try to outline some key principles.

1. **A project needs to have an *aim*.**

   Carrying out computations on a given structure, a given chemical reaction, or a given set of atomic motions is not in itself of huge scientific value. It acquires that value through helping to test a theory, or seeking to predict a new property, or trying to come up with an idea for improving an existing set of reaction conditions. Sometimes, it can be fruitful to start out on a computational study without explicitly articulating what you are trying to find out. Initial results can help to narrow down the intuitively grasped objectives and thereby to define them after the fact. But it is very rare that you can bypass this step altogether.

2. **You must choose the right method and the right computational implementation for the problem at hand.**

   There are many different techniques available, but they have different scopes and different strengths. Choosing between a molecular mechanical and a quantum chemical approach, for example, is a big decision to make and you can save a lot of time by making that choice correctly. While the various computer programs available for doing quantum chemical calculations are quite similar in some respects, they differ widely in others, and the right code can really make the difference between full success and complete failure.

3. **You must choose a *model* carefully.**

   Computation, at least in the sense covered in this book, always involves the choice of an atomistic model, a microscopic system of atoms and molecules whose modelled behaviour is held to be informative about the macroscopic behaviour that you wish to understand. It is easy to leave out some key 'ingredient' without which the model is simply inadequate. For example, if an impurity or a defect plays a major role, then a model containing only the major species, or the perfect crystal, will simply not deliver results that will be right for the right reason. Conversely, if the point you wish to explore can be described simply, then choosing a very elaborate model to do so, just because this is possible computationally, may represent a waste of the researcher's own (and the computer's) time—and may even obscure the simple answer that was being sought.

4. **It is important to be critical of computational results.**

   Most or indeed all calculations require severe approximations, and thereby yield results that are objectively speaking *wrong*. That does not mean that within the accuracy they achieve they cannot be useful. ('All models are wrong—but some are useful', as was once said by the British statistician George Box.) But it does invite caution. Just like a stopped clock shows the right time twice a day, severely flawed calculations can yield apparently impressive agreement with experiment in some or other way. This can lead you to unwarily grant them much more respect than they deserve for their predictions concerning another aspect, where their performance cannot be so readily assessed. One way to avoid this problem is to do the best calculations you can for the problem at hand. Often there is a reasonably obvious hierarchy of quality for the various approximate techniques available, though even this hierarchy is sadly much less clear than you might like.

Another way to avoid this problem is important enough to be separated out into a guideline in its own right.

5. **The best way to avoid fooling yourself in this way is to *test* the calculations.**

   In many well-performed projects, the proportion of the total number of calculations that end up being remembered and published in the literature is very small, ten per cent or less. Some of these will be tests. Of the remaining 90%, many will be additional tests, and some will be calculations rejected based on yet other tests. If measurements are available for a property or a system that is similar to the one that you are studying, then you should carry out the calculations on that system or that property as well as the one that is the main focus of the work. Or you should check the literature to see if this test has been performed by someone else elsewhere. Such tests are also really helpful for learning about what a given computational method actually does: using it to look at a system for which you are fairly confident that you understand both the theoretical background and the experimental observations can be really educational. Sometimes it can also help to understand the theory. Not by accident, many of the best test calculations are also very quick to perform: short simple tasks, where both the input the user must generate and the output he or she must digest are brief and to the point, are a very useful proving ground before turning to calculations that take weeks to prepare, weeks to run, and months to analyse.

9.3 Summary

- Computational methods now play an essential role in many areas of chemistry.
- The broad range of methods and software available means that it is easier than ever to tackle computational projects.
- Careful design of a project requires that the aims be well defined, and the uncertainties well understood.

Index